Command and Control:
ICS, Strategy Development and Tactical Selections

Robert Murgallis
Burton W. Phelps

edited by John Joerschke

Published by
Fire Protection Publications • Oklahoma State University
Stillwater, Oklahoma

Project Manager: Cindy Pickering
Design/Layout: Ann Moffat
Graphics: Ann Moffat and Ben Brock

Photographs courtesy of the following:
Edward Burns: Figures 6.3 and 8.1
Don Forsyth: Figure 5.6
Katherine Kern Murgallis: Original concept for Figure 11.3 and scales on page 55
Robert Murgallis: Figures 1.1, 3.16, 4.1, 5.1, 5.2, 5.4, 6.1, 6.2, .71, 7.2, 9.1 – 9.4, 10.2, 10.4 – 10.7, 11.1, 11.4, and 11.5
Burton Phelps: Figures 2.1, 3.7, 3.18, 4.2 – 4.4, 5.3, 5.5, 8.2, 8.3, 10.1, 10.3, and 11.2

ISBN 0-87939-199-5
Library of Congress Control Number: 2001097419
First Edition
First Printing, January 2002

Printed in the United States of America

Contents

About the Authors

Robert Murgallis is a 31-year veteran of the fire service, having retired as a deputy chief of the City of Santa Clara, California, Fire Department. He is a graduate of the National Fire Academy's Executive Fire Officer Program, and holds an associate of fire science degree and a bachelor of science degree in public administration. A fire service instructor for more than 25 years, including 15 years as an adjunct faculty member and course developer at the National Fire Academy, Chief Murgallis is the owner of Custom Seminars, a consulting firm specializing in custom course design and assessment lab development for the fire service. He has made several presentations in the areas of training, management, and operations at fire service conferences. Chief Murgallis has written numerous articles for fire service trade publications and was a contributing author for *Model Procedures Guides on High-Rise Fire Fighting and Hazardous Materials Response.*

Burton W. Phelps is a 30-year veteran of the fire service, having retired as a deputy chief of the Anne Arundel County, Maryland, Fire Department. He holds an associate of fire science degree and a bachelor of science management. A fire service instructor for more than 32 years, including 22 years as an adjunct faculty member and course developer at the National Fire Academy, Chief Phelps is the owner of Burton W. Phelps, Inc., a consulting firm specializing in ICS and strategy-tactics course design for the fire service. He is also president of Innovative Computer Simulations, Inc., a fire-service-training software company. He has made several presentations in the areas of training and operations at fire service conferences. Chief Phelps has written numerous articles for fire service trade publications and was a contributing author for *Model Procedures Guides on Structural Firefighting, High-Rise Fire Fighting and Hazardous Materials Response.*

Acknowledgements

First and foremost we would like to thank our wives, Katherine Murgallis and Tonna Phelps, and our families. Their support and encouragement created an atmosphere of success.

To all those peers and mentors who guided, encouraged, and cajoled us through our careers, enabling us to arrive at where we are today. This would most certainly include that elite cadre of fire service professionals and national experts who continue to serve, sharing a wealth of information and experience, as National Fire Academy Instructors.

To all of our brothers and sisters in the fire service, both those who continue to fight the fight and those who have paid the ultimate price. Never forget yours is among the noblest of professions and your mission of utmost importance.

A special thanks to Assistant Chief Edward Burns, Ret., Los Angeles City Fire Department and Battalion Chief Edward McDonald, Ret., Fire Department New York. Their friendship, guidance, and assistance has been invaluable.

Chapter 1
Introduction

Chapter 1

Introduction

You are a chief officer, and you have arrived at a fire involving a vacant commercial structure (Figure 1.1). The building is two stories, 100 feet by 75 feet. There is a 25 percent involvement on the first floor from the center of the building to the front wall. Your preplan indicates that there are two open stairways, one in the center of the building and one against the rear wall. The front has a set of double doors, and there is a side door at the rear of the right-hand wall. You have three engines, one ladder truck, and a chief on the assignment. Now it is time to earn that promotion to chief. Everything that you have learned will be on the line. Will you determine the right strategy and tactics and correctly place the apparatus and personnel?

Tens of thousands of fire incidents occur every day. Thousands of company and chief officers face these types of decisions. Not surprisingly,

Figure 1.1. Vacant storefront with severe exposures.

many officers get it wrong. Often they may not even consider strategy and tactics, and even if they do, they select the wrong ones. They take unnecessary risks, place people and apparatus in the wrong locations, allow the fire to spread, and then take a ribbing from their peers and subordinates. In the really bad cases, they find themselves in court answering charges of neglect or downright incompetence.

How can you learn the correct course of action? Perhaps you are in a small department and face few actual fires, so you can't get the real experience you need. Or you are in a moderate size department, but working fire incidents are few and far between. You still can't get the real experience. Or you are in a large department, but the officers you are using as mentors are from the old school, and you believe that you are learning bad habits. Once again, you can't get the proper experience. Where do you go, and what do you do?

Nothing can take the place of actual experience at an incident. But for that experience to be truly valuable, you must be getting your education from an expert. This is a fire officer who has been around the block—one who has been to a large number of fires and also has taken training throughout his or her career. The fires must have been one-alarm or two-alarm or greater incidents, working fires where crews were operating for longer periods of time. Just riding in the fire engine to a false alarm or to burning food on the stove does not count. Experience is the key to effective decision making on the incident scene.

But you probably are not in a department where the number of fires is increasing. Where do you go to get real, meaningful experience? Well, you take training classes in strategy and tactics, the Incident Command System, water supply, specific occupancies (for example, strip shopping centers and garden apartments), and command and control. You read the fire journals, textbooks, and training books such as this.

After your training, you do simulations. You apply under pressure what you have learned in the class or from a book. This is how you get your experience. It isn't easy and it doesn't happen overnight. But, with time, you gain experience. Do not take that need for time lightly. It is the most critical element to becoming an experienced officer. Remember, you have not learned any subject until you have applied your theoretical knowledge in real-world terms (simulations) many times over.

How does this book figure into your plans for career experience? It explains the general processes you need to learn for controlling all incidents, and it details some of the specific information you will need in order to make effective decisions at certain common occupancies. These things you need to learn to be on your way as an experienced fire officer.

In this book, we cover Incident Scene Decision Making in depth. You will learn the two presently used methodologies for successful decision making under emergency conditions. You will learn how to develop your memory in order to make lightning-fast correct decisions based on your knowledge of various types of occupancies. And you will learn how to develop effective strategies and tactics.

We cover the basic elements of the Incident Command System—not the whole, confusing ICS, but just those functions that you will need in order to handle the vast majority of incidents to which you respond.

Then we cover specific information about garden apartments, strip shopping centers, vacant commercial buildings, lumber yards, small hotels and boarding houses, and enclosed malls. In these chapters we get down to the specifics and explain:

- Occupancy specific cues.
- Cue-based predictions.
- Incident management cues.
- Strategy and tactics.

We consider safety issues at the incident scene, resource allocation, and personnel accountability. We also discuss resource availability, resource allocation, and other incident support issues.

This book will provide you with the foundation for effective decision making, as well as a great deal of information on specific types of occupancies. By studying and completing this text, you will be on your way to becoming an experienced officer. Good luck!

Oh, we almost forgot. What about the answer to the scenario at the beginning of this chapter? Here it is.

Size-up

The building is 25 percent involved from the center to the front doors. The building is vacant. There has been possible deterioration of the building floors and stairs. Something to warn personnel about. There may be vagrants sleeping or living in the building. A possible life hazard. This may be an arson fire. Building has open stair shafts.

Problem Identification

- Fire on first floor
- Possible vagrants—life hazard
- Possible spread to second floor through open stairs

Water Supply Needs (NFA Fire-Flow Formula)

- 600 gpm on first floor
- 150 gpm to check extension on second floor

Strategy (First-In Officer and Chief)

- Rescue (primary search)
- Confinement
- Extinguishment
- Ventilation

Tactics

- Attack fire from door on right-hand wall and push the fire from the unburned to the burned area using a master stream device set inside the door—600 gpm.
- Advance a hoseline up the rear stairs to the second floor to check for extension—150 gpm.
- Ventilate horizontally on the first and second floors.
- If there are reports of too much fire or deteriorated conditions inside, especially the stairs, move to a defensive operation.

Isn't it simple when you see someone else do it? At the completion of this book, you should be on your way to developing and implementing such successful operations.

Chapter 2
Incident Scene Decision Making

Incident Scene Decision Making

Chapter 2

Overview of Incident Scene Decision Making

You are an engine-company captain standing in front of a three-story garden apartment (Figure 2.1). The building has four apartments per floor, with one of the ground-floor apartment spaces being used as a storage/laundry room for the occupants. There is heavy fire in the storage room, about a 50 percent involvement of a space that is 40 feet by 21 feet. Heavy smoke is apparent on the second floor, especially above the storage/laundry room area. The time is 0500 hours.

What orders/directives are you going to give to your personnel and other responding units? How are you going to make those decisions? Incident-scene decision makers use two primary methods to reach conclu-

Figure 2.1. Three-story garden apartment fire on arrival.

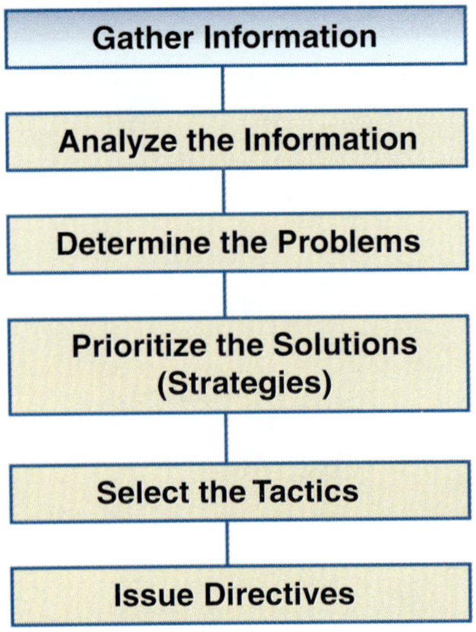

Figure 2.2. Classical decision making.

sions, determine results, and institute actions. They are the **classical method** and the **recognition-primed method** (Figures 2.2 and 2.3).

The classical method is a time-consuming process in which the decision maker:

1. Gathers information
2. Analyzes the information
3. Determines the problems that are present and selects and prioritizes those problems
4. Determines and prioritizes the possible solutions
5. Selects tactics from one or more possible options
6. Issues directives to have the tactics implemented

The recognition-primed method is a more rapid process in which the decision maker:

1. Looks for certain critical cues (visual, verbal, touch, smell)
2. Relates those cues to previous similar situations (from experience or training)
3. Recalls the previous conclusion, results, and actions that most fit the new situation
4. Issues directives to have the tactics implemented

Which methodology can you use? The recognition-primed is better by far. It is a method of short-cutting the long process described in the classical methodology. But can you use it? That depends on a number of factors. Do you understand the two methods and how they interrelate? What are your previous experiences and training? How many of these situations have you experienced before? One, two, ten, twenty?

First, let's look at recognition-primed decision making. Where did this methodology originate? Finding its capability declining due to inexperienced wartime commanders, the United States Army funded a study in the

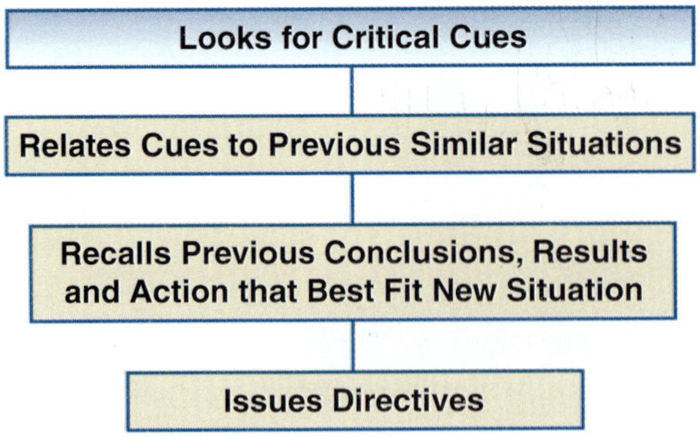

Figure 2.3. Recognition-primed decision making.

early 1990s. The study's objective was to examine how highly proficient personnel make decisions under conditions of extreme time pressure and how those decisions could affect lives and property. Klein and Associates, an Ohio firm, was awarded the contract to do the study.

Because no wars were occurring anywhere in the world where U.S. troops were operating, the study examined the fire service, paying specific attention to the incident commanders and their decisions during specific incidents. To do this, Klein and Associates interviewed 26 experienced incident commanders. The results were as follows. In fewer than 12 percent of the cases was there any evidence of simultaneous comparisons and relative evaluation of two or more options. In other words, the experienced incident commanders almost never followed the classical method or considered more than one tactical option to achieve a specific result.

In more than 80 percent of the cases, the incident commanders used their experience to directly identify the situation as typical of a standard prototype and to identify a typical course of action for that prototype. In this way, the ICs handled decisions without any need to consider more than one option. The recognition-primed model was synthesized from these research data, which emphasized the use of recognition rather than calculation or analysis for rapid decision making. This concept was brought to the National Fire Academy and applied to training programs and scenarios relating to Fire Service incident management by Hugh Wood, former Program Chair for Incident Management Programs.

What is a prototype? It is a previously encountered situation, or a synthesis of many situations, similar enough to a current situation to indicate effective courses of action. To put it simply, "I've seen this before; this is what I did before; I'll do something similar at this incident." Success! When the incident commanders encountered a decision point, they recognized a match to a mental prototype. By applying their experience to the prototype, they determined how to proceed in the situation at hand.

The more experience you have on the same type of incident, the better you will be able to read the subtle differences at such incidents, draw refined conclusions, and direct the most appropriate actions to provide a solution. Obviously, basing decisions on understanding from previous experience can produce action results much faster than following the step-by-step classical process.

Normally, we all try to use a recognition-primed methodology whenever we have to respond to any challenge that needs a response. We immediately search our brain for an answer. What we are really doing is trying to relate the problem to something that we have seen before and to recall the previous answer. If we remember a previous situation that we responded to successfully, then we simply respond in a similar manner.

However, what if we can recall no previous situation that we responded to successfully? We have two choices—panic or think it through. Too often, fire officers who have limited experience tend to "lock-up," or panic. They freeze and cannot respond appropriately to the problem, and then the outcome is bad.

Let's look at the use of the two methodologies to understand what they are all about and when it is appropriate to use them.

Appropriate Use of Decision-Making Methods

First, let's look at the word *cue*. A cue is a hint, a stimulus that guides behavior, a signal to begin. For us in the fire service, a cue is simply a piece of critical information upon which we base our decisions. In this context, a cue is a sensory observation, usually visual or verbal, which by itself or in conjunction with other cues, leads a decision maker to relate to previously encountered situations (experience or training). This leads in turn to recalling the previous conclusions, results, or actions. For example, if the cue is fire and smoke location in a building, you will search your memories of similar situations to determine the best direction of attack. Or if the cue is the type of occupancy, you will search your memory to decide if it is a high life hazard. Or if the cue is the type of roof construction, you will search for memories that help determine whether you can commit firefighters to the roof.

The Classical Method

Decision makers must rely on the classical method when they are in the training mode. In the training mode, they will be taught to look for critical cues for different types of incidents, draw correct conclusions, consider effective results, and take action for the particular type of incident being studied. Whether the students have learned the cues, conclusions, results, and actions must be tested in an application format. Such a format is a simulation. Because of the dynamics of incident scene operations, training on incident-scene decision making is not complete until the students have been tested under conditions similar to those at an actual incident scene. Just sitting in a classroom and being lectured is only part of the learning process. You have not learned the skills until you have practiced them under the conditions in which you will actually use them. You must receive such training from an expert in the incident type: for example, an urban or city fire officer should learn wildland fire fighting from a wildland expert.

The classical process is used for evaluating and planning when time is not a factor. When fire officers do community risk assessment and develop plans for various types of operations, we use the classical method. Why? Because most of us have little actual experience with the types of incidents for which we are writing operational plans—earthquakes, tornadoes, floods, hurricanes, man-made disasters. Or how about our prefire plans at the company level? We go to an establishment, gather information, take measurements, and return to quarters. At quarters, we gather the crew around the table and discuss the various possibilities, building construction, and problems they likely will encounter under fire conditions. Again, personnel apply the classical method as they process the information.

Decision makers employ the classical method when they evaluate and compare other decision makers' use of critical cues to draw conclusions, determine desired results, and take appropriate actions. This form of training typically involves case studies. The student uses a case study to examine both the obvious and the subtle differences in cues. This provides optional conclusions, results, and action-sets based on those differences. By using a case study and the classical method, students can evaluate whether the conclusions and actions of the decision maker at the actual scene were appropriate for the cues. If they were not, then the students may learn specific actions to avoid.

In addition, you will use the classical method at an actual incident scene where you have little or no training in or experience with the specific type of incident. In this situation, you already will have tried to use the recognition-primed method. Failing to recall any prototype situation or successful actions, then you will be ready either to panic or to work your way through the incident. The better choice, of course, is to work your way through the incident using the classical methodology.

As a decision maker you must form a basic plan before directing tactical actions. Any course of action that does not include an evaluation of the incident information, a risk-benefit analysis, and appropriate strategies and tactics is not a plan—it is a design for disaster. Base your plan on incident information (critical cues), real problems, and appropriate broad solutions (strategies). Then choose the best solutions (tactics) from your available options.

Again, the steps of the classical methodology (Command Sequence) are:

1. Gather information.
2. Analyze the information.
3. Determine the problems that are present and select and prioritize those problems.
4. Determine and prioritize the possible solutions (strategies).
5. Select tactics from one or more possible options.
6. Issue directives to implement the tactics.

The Recognition-Primed Method

You will use the recognition-primed method when you have adequate experience with or training in the incident type or the particular variables present. The recognition-primed method is almost instant recall of previously learned conclusions, results, and actions. It includes the interrelationships of specific information with conclusions, results, and actions based on whether they worked before. Therefore, it provides a direct, lightning-fast transition from what you see, hear, feel, and smell to what you conclude and what you do.

Because of the time pressures in emergency-scene decision making, the choice between the recognition-primed method and the classical method will not be conscious. As a decision maker, you will always attempt the recognition-primed method first, and you must recognize when you possess insufficient information to use it. Some cues that your information is not adequate are:

- It is obvious to you that you have little or no experience with or training in the specific incident type.
- You recognize that the incident cues are very unfamiliar and do not immediately lead to appropriate action decisions.
- You feel lost or overwhelmed, cannot think, or are in a panic.

In these cases, the classical method is the appropriate response.

How do you learn to use the recognition-primed methodology? First, it is an experience-based methodology. You need to gain experience in various types of incidents. And how do you do that? You respond to as many incidents as possible, learning all you can from each and every one of them. Of course, in a small to moderate sized community with a limited number of incidents, you could spend a lifetime trying to gain that experience. Also, if the officers from whom you are attempting to learn the correct actions for a situation are themselves responding incorrectly, you will merely learn to make very rapid wrong decisions.

Other effective ways of learning are also available. Take training from subject matter experts. Find those people in your department or neighboring departments who can provide expert training on different occupancy types. Take their training classes, but make sure that you follow up with simulation training on those occupancy-types. You must develop your knowledge-based decision-making ability in various situations under pressures like those at an actual incident scene.

The Command Sequence

The command sequence is another name for the classical methodology of decision making. To attain the expertise of an effective incident commander, you must receive effective training and gain experience analyzing various incident types. The Command Sequence is part of the process that will move you from novice to expert.

Table 2.1. The Command Sequence

Action	Result
Size-up	Problem Identification
Strategy and Tactics	Action Plan
Implementing the Action Plan	Tasks

As stated before, you will have to use the Command Sequence at any incident where you do not have the training or experience for recognition-primed decision making. In your career, you will come across numerous situations where you lack training and experience and must rely on classical decision making.

The command sequence emanates from the mission of the fire department, and its incident priorities are the same as those in the fire department's mission statement:

1. Life safety
2. Incident stabilization
3. Property conservation

The mission statement describes why a fire department exists. Every action the department takes must address one or more of these goals.

The command sequence involves three "action steps," each aimed at a specific result (Table 2.1). It is a cyclical process in which, as each action step achieves its intended result, that result leads in turn to the next action step. The first action step, size-up, results in identifying the problem; identifying the problem facilitates the second action step, determining strategy and tactics; strategy and tactics then result in an action plan; and the action step, implementing the action plan, results in assigned tasks; finally, the effectiveness of those assigned tasks must be evaluated in an ongoing size-up, which begins the entire cycle anew.

Size-Up

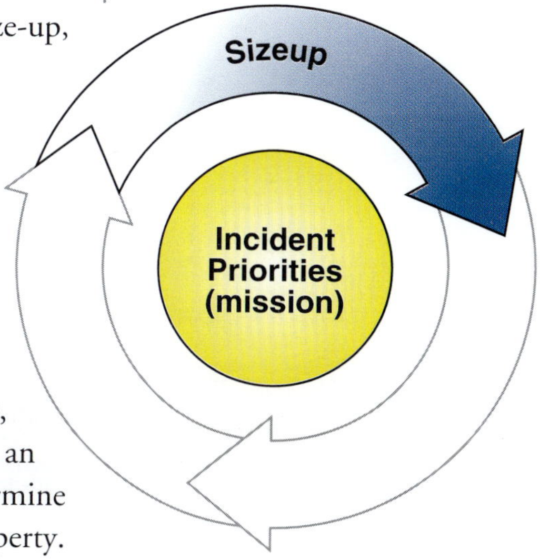

Size-up is the process of gathering and analyzing incident information that affects our decision making. We call this type of information critical cues. Knowing where to look for these critical cues enhances our ability to identify the problems at the incident. The key to this process is to have the knowledge and experience to know what cues to look for and where to look for them. For the incident commander, the most important step is to complete a correct size-up. Without an appropriate and accurate size-up, it is impossible even to begin to determine the solutions needed that will prevent further damage to life and property.

The three phases of size-up are:

1. Preincident information
2. Dispatch through on-scene size-up
3. Ongoing size-up

Phase One: Preincident Information. Size-up is more than what you see through the windshield when you arrive at the scene. It includes preincident information, the factors (critical cues) that are known or gathered before the incident. These factors will affect decisions and actions. Information is available on every structure or facility, and you can obtain it simply by asking questions and taking the time to gather it. This information includes:

Size-Up

Pre-Incident

- Preplan documents
  - — Occupancy information
  - — Building construction
  - — Hazards to personnel
  - — Fire-flow needs
  - — Fire behavior probability
  - — Water supply location and quantity
  - — Problems anticipated
  - — Built-in fire protection
  - — Floor plan/plot plan
  - — Utility connections/shutoffs
- Environmental considerations
- Time of day, day of week, season of year
- Knowledge of the surrounding area
- Departmental resources
- Interagency/private sector assistance

Preincident information helps the command officer understand what has happened. It will assist in determining what is happening presently, as well as in predicting what is going to happen. It provides a basis for being proactive. It is critical to the identification of the problems, as well as the development of strategy and tactics. It is invaluable in identifying safety considerations, including any hazards present. As many chiefs have often said, "It is hard to gather preincident information when the flames are licking at the boxes of dynamite." Preincident information must be gathered ahead of time—before there is an incident.

Every fire department needs to implement a prefire planning program. There is no excuse for not having a system that increases the knowledge of the responding fire officers and firefighters. Without an effective prefire planning program, you will be doing a lot of guessing in your decisions about the lives and property of your citizens. We all like to call ourselves professionals, but are we really professional if we ignore critical programs that will be of utmost benefit to our customers?

Phase Two: Dispatch Through On-Scene Size-up. While preincident information from phase one provides a foundation for the more specific data needed to understand a particular incident, phase two is a rapid gathering and evaluation of factors (critical cues) related to a specific emergency incident. Identifying the problems is phase two's first objective. The command officer must assess incident conditions. He or she must know where to look and what to look for and must understand the significance of the critical cues. If we do not identify the real problems, we probably will apply the wrong solutions. Potential hazards to firefighters need to be "read" and identified.

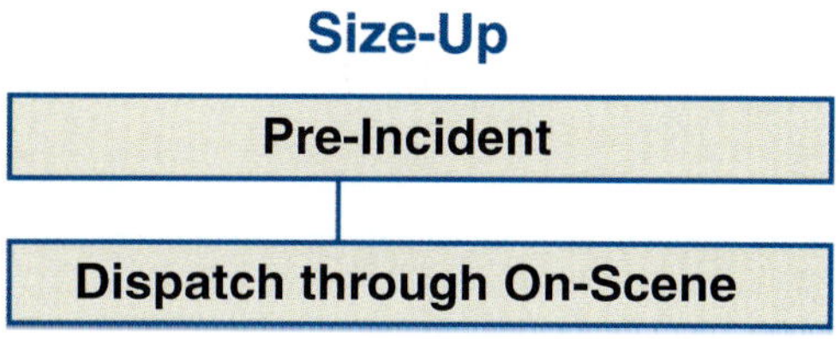

When relieving an initial IC, the command officer must review the big picture, evaluate resource needs, the current strategy and tactics, changes in incident conditions, and the probability of escalation.

Size-Up Considerations. Various authors list different elements to delineate the factors that affect decision making. One technique lists 13 points using the acronym WALLACE WAS HOT:

Water

Area

Life

Location/Extent

Apparatus/Personnel

Construction

Exposures

Weather

Auxiliary appliances (standpipes, sprinklers, heat detectors, etc.)

Special hazards

Height

Occupancy

Time

These 13 points are specific enough to encompass emergency incident scene factors, yet broad enough to fit almost any type of incident.

Once you have learned it, the acronym will help you to recall these topics. However, most people find that it is impossible to process so many items en route to or standing in front of an incident scene. The most appropriate uses of WALLACE WAS HOT are for training—especially training that involves learning the critical cues for handling a specific incident type—and for prefire planning with the exception of location/extent, weather, and time. Every fire officer and hopeful fire officer should know the size-up factors and be able to recite them from memory. How can you discuss incident problems and strategy and tactics if you don't know the critical types of information that must be gathered?

Size-Up Information Sources. Dispatchers provide valuable incident information. Train your dispatch personnel to give responding companies all the important information. Your knowledge base (what you know about the area, structure, water supply, etc.), includes information from people on scene, including fire and police personnel and civilians. Do not overlook any source of information concerning an incident. Be open-minded, ready to listen to others, and prepared to make decisions based on your best judgment of the data.

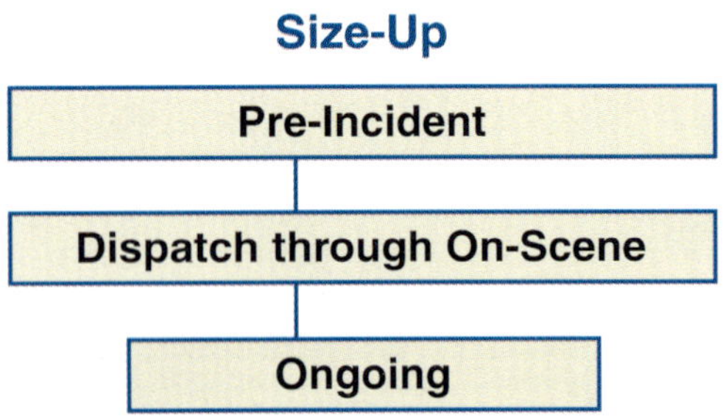

Phase Three: Ongoing Size-Up. Incidents are dynamic, and the IC must continuously evaluate the situation. Due to constantly changing conditions, new problems may arise as previously identified problems are solved. The IC must evaluate actions taken and continually assess their results, asking herself or himself, "Are our actions solving the problems at this incident scene? Are our tactics working?" A reevaluation of the critical cues may indicate that a modification of the action plan is needed.

Size-up must be a continuous process—from the first arriving officer to the last officer who leaves the scene. If you are not thinking about the situation, how it is changing, what new actions need to be done now, what might occur, and what steps you might need to take, then you are not doing the job you have been promoted to do. Both during and after the battle, you need to be constantly mentally involved with the incident.

Command officers must view an incident scene as pessimists, people who continually look for things that could go wrong, and are therefore better prepared to deal with them. Remember, Murphy's Law applies here also!

Problem Identification

The goal of size-up is problem identification. Mental analysis and comparison of incident cues is the basis for problem identification. Once you have identified the problems at an incident, you may determine and apply the correct solutions. Now the planning process can begin.

Take, for example, the garden apartment fire described at the beginning of this chapter. What problems can we identify?

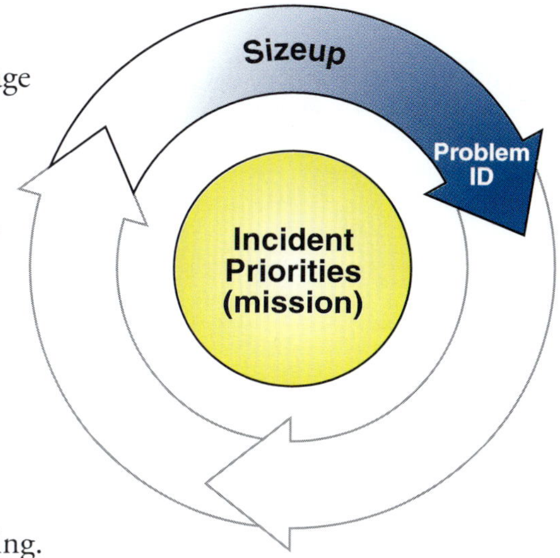

- Fire in the ground-level storage room at a 50 percent level.
- Heavy smoke on the second floor, especially over the storage room.
- Probably lighter smoke on the third floor.

These are the obvious ones. Now what else can you predict from your base of information about residential, multistory dwellings?

- At 0500 hours most people will still be asleep; major rescue problem.
- Fire spread in open storage room will be rapid.
- Ventilating the storage room will be difficult.
- Forcing steel-clad doors to apartments will be time-consuming.
- Searching in heavy smoke conditions will be slow; people may perish.
- Fire spread to floor above will be enhanced by the open ceiling area in storage room.
- Fire may enter walls of floor above and extend upward.
- Situation requires multiple alarms of resources.
- Situation requires a heavy EMS response.

Prediction is a critical component of size-up and a skill that you must develop if you are to be an effective incident commander. Experienced fire officers can make predictions concerning any occupancy if they have had training in or real experience with that occupancy type. As you can see, it is critical to read more than the simple facts that stare you in the face at an incident. How would our decision making be affected if we failed to predict the second set of not so obvious critical cues?

Strategy and Tactics

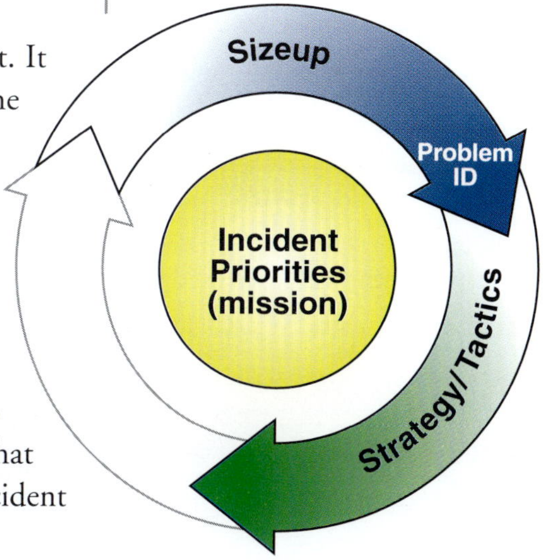

Strategy is the overall plan that will be used to control the incident. It delineates the broad goals, defines "what" must be done to solve the problems, and provides the basis for action planning. Tactics, on the other hand, are the specific operations that must be completed successfully to accomplish the strategy's goals. They are the actions that the IC is going to take.

Determining Strategy. Strategy evolves directly from the identified problems, and it is the beginning of the solution to those problems. It is part of the "think" process, as opposed to the "act process." No actions are taken at the strategy level. It is the part of the process that allows you to prioritize the actions that must be taken to bring the incident

to a successful conclusion. Strategy gives directions for getting from where you are to where you want to be. It may have multiple components in order to gain control of an incident.

Generally, command officers use Lloyd Layman's seven factors as a basis for developing strategy:

Rescue
Exposure
Confinement
Extinguishment
Overhaul
Ventilation
Salvage

Remember these factors by using the acronym RECEO VS (Ree Cee Ooh Vee Ess). The first five factors, RECEO, are in priority order. Typically, ventilation supports one or more of the other strategies. As the incident commander at the beginning of a solution, you probably will not immediately concern yourself with overhaul and salvage. You or a relieving officer will address those items later as the event progresses. Initially, you must concentrate on the effort needed to bring the situation under control.

Here's how RECEO VS works. Once the problems are identified, the incident commander determines which of the strategy elements apply to the situation. For example, let's say that we are at an automobile fire on a parking lot. No one is in or near the car when you arrive with your engine company. A quick evaluation of the situation would determine that no one was in any danger of being injured by the event. There is no rescue problem; therefore, there is no need for a rescue strategy. You can move on to exposures, confinement, and extinguishment as necessary.

Or consider the garden apartment incident. A quick evaluation obviously will include the fact that people are probably trapped in the structure due to the smoke conditions. Now rescue is part of the solution, and you must have a rescue strategy.

Since rescue is the highest priority, when can you move on to protecting exposures? That is a straightforward question, but many fire officers are not sure of the answer, although it is rather simple. When the incident commander has addressed the rescue strategy with sufficient tactics and resources to effectively begin the solution to the rescue problem, then the IC can move on to exposure protection or a lower-priority strategy. However, if the people needing rescue cannot be rescued due to location or danger to firefighters and action has been taken to protect them as well as possible, then the IC may move on to exposure protection or a lower-priority strategy.

A well-defined strategy gives incident personnel a clear description of the IC's plan and helps them accomplish it. The IC must determine strategy before developing an action plan. Having a strategy indicates that critical cues have been gathered and assessed and problems have been identified. The IC also has completed an evaluation of resource requirements and availability and has set priorities. Planning has begun.

Selecting Tactics. Once you have determined an appropriate strategy, you must select the tactics needed to carry out your strategy. The tactics phase is still part of the thinking phase of the process. Until directives are issued to companies and crews, no action will take place.

When you use the recognition-primed methodology, you will select tactics from your knowledge base. In other words, if you haven't seen a tactic, trained on it, or used it before, you probably cannot select it from your memory. Experienced incident commanders have gathered a large number of possible tactical operations into their memories. They are usually avid readers of fire journals, take numerous training classes to stay abreast of new concepts, and pay attention to the decisions of others when they are not in command.

Until you are experienced in incident scene tactical operations, you may develop tactics from your analysis of the situation. You have probably faced situations where you rejected all of the tactics that you knew from experience and tried another action that you hoped would be effective. Experienced fire officers also do this at incidents with which they have limited or no experience. For example, a child falls from a wall into a gorilla pen at a zoo. You must decide what to do by using the classical methodology—unless of course, you have experience in such situations.

Some selected tactics and the strategies they address are:

- Confine fire to room of origin (confinement strategy).
- Conduct a primary search (rescue strategy).
- Provide horizontal ventilation (rescue and confinement strategies).
- Check for extension (confinement strategy).

Tactics based on appropriate strategies will guide the operational performance required of the companies or crews.

Let's add another variable to the garden apartment scenario. The fire is exiting the storage room door and threatening the only stairway that leads to the second and third floors, thereby jeopardizing the firefighters' access to those floors and compromising the occupants' exit. The IC orders a line placed at the storage-room door to push the fire back into the storage room.

This is a good tactic because it will protect the stairs and allow for the movement of firefighters and citizens in the stairway.

Is this a confinement tactic or a rescue tactic? To determine what strategy is being achieved primarily, you must consider the primary reason for employing the tactic. In this instance, the IC quite obviously is trying to allow firefighters to reach rescue positions on the second and third floors and to allow citizens a means of effecting their own escape, if they are able. While partial confinement of the fire is also being accomplished, it is obviously not the primary reason for the tactical operation. Therefore, this is a rescue tactic. If the rescue were completed and the line continued to operate at pushing the fire back into the storage room, then the operation would become a confinement tactic. Yes, things are dynamic on the incident scene.

Tactics are the "how" of the solutions to the problems. They are measurable and specific and can be completed within a designated time. The IC prioritizes the order of tactical activities according to the priorities of the strategies they serve.

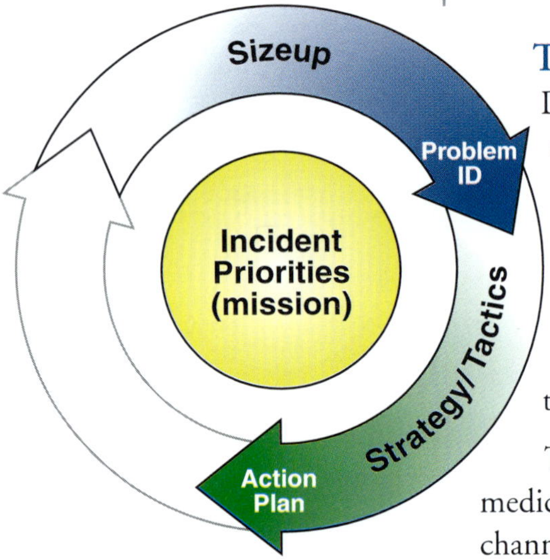

The Action Plan

Determining strategies and selecting tactics enables the command officer to develop an action plan. In addition to strategy (the "what" of incident solutions) and tactics (the "how"), the action plan must also describe the "who," the "where," and the "when." The action plan defines the ICS organizational structure for the operating forces, and the resource assignments within the organizational structure establish the "who." Adequate resources must be assigned for tactics to be completed successfully.

The action plan must also include a communication plan and a medical plan. The communication plan must define the operational channels for the incident. For normal incidents, most departments use a constant channel or channel set, and the communication plan is a habit. However, when incidents evolve into major situations, the small-incident communication plan may need modification. The medical plan must ensure that emergency medical services are available and on scene to care for injured response personnel and injured civilians.

Implementation and Tasks

Once an action plan is in place, the command officer can determine how best to implement it and then begin assigning the necessary tasks. Resources need to know their role in the plan. Directives issued to those resources delineate their particular "whats," "whens," and "wheres." These directives can describe either the overall tactical operation or the specific tasks to be accomplished. A tactical order would simply say, "Truck 2, perform vertical and horizontal ventilation on arrival." This order would leave much of the decision making to the company officer. A task order, on the other hand, would say, "Truck 2, ladder the rear of the building and cut a 4-foot by 4-foot hole in the C-D corner." This directive delineates the specific tasks and allows the company officer to do exactly what he or she is told and nothing more.

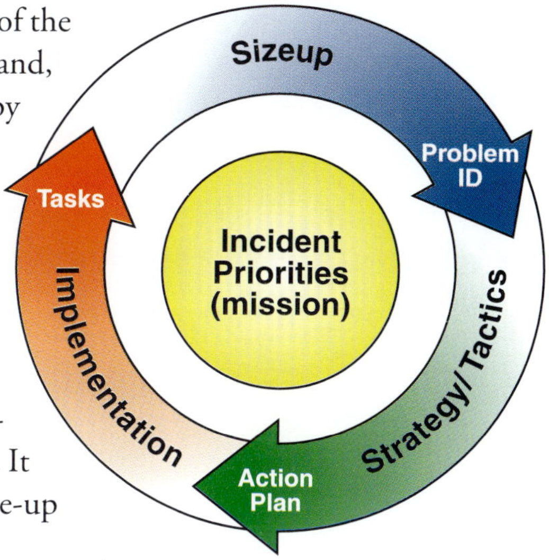

Performing the tasks required for tactical operations is the final result in each cycle of the command sequence. Crews perform specially learned manual tasks that, when completed, achieve or help to achieve a tactic. As tactical operations are performed, their evaluation cycles back into the first step in the command sequence, size-up. It is absolutely critical that the IC continues to perform ongoing size-up throughout the incident.

Conclusion

Following the command sequence process helps to ensure that you will not overlook critical problems, and solutions. Because it makes you think before you act, it helps to keep you in a proactive mode and to get you ahead of the time curve. And because it outlines a logical thought process, you can use it at any type of incident.

Activity 2.1
Use of Size-Up to Identify Problems

Purpose

To identify the appropriate critical cues and the incident problems that those cues indicate. This activity may also be done with a group of personnel. If a group works on the same problem, they are likely list a more complete set of cues than would one person.

Directions

1. Read the scenario of a fire situation and examine the accompanying floor plan. You will assume the role of the first-in officer at the incident. You should be able to complete this activity in 10 minutes.
2. Complete the lists on the next page.
   a. Specify the critical cues.
   b. Identify the incident problems.

Scenario

You are the first fire department officer to arrive at a fire involving a strip shopping center. The shopping center is 120-feet long by 60-feet deep. There are five different occupancies in the strip shopping center, all of equal size (24 feet by 60 feet). The building is of ordinary (masonry, wood-joist) construction. The floor is a concrete slab. The roof assembly is ridge-pole and rafter construction and is common to all five occupancies. There is a false mansard on the front of the building. The fire is in the second occupancy from the left. There is a 50 percent involvement in the occupancy. Smoke is issuing from attic vents and from under the eves. There is heavy smoke in adjoining occupancies. There is no sprinkler system. There are no other exposed buildings near the strip shopping center. It is 0245 hours.

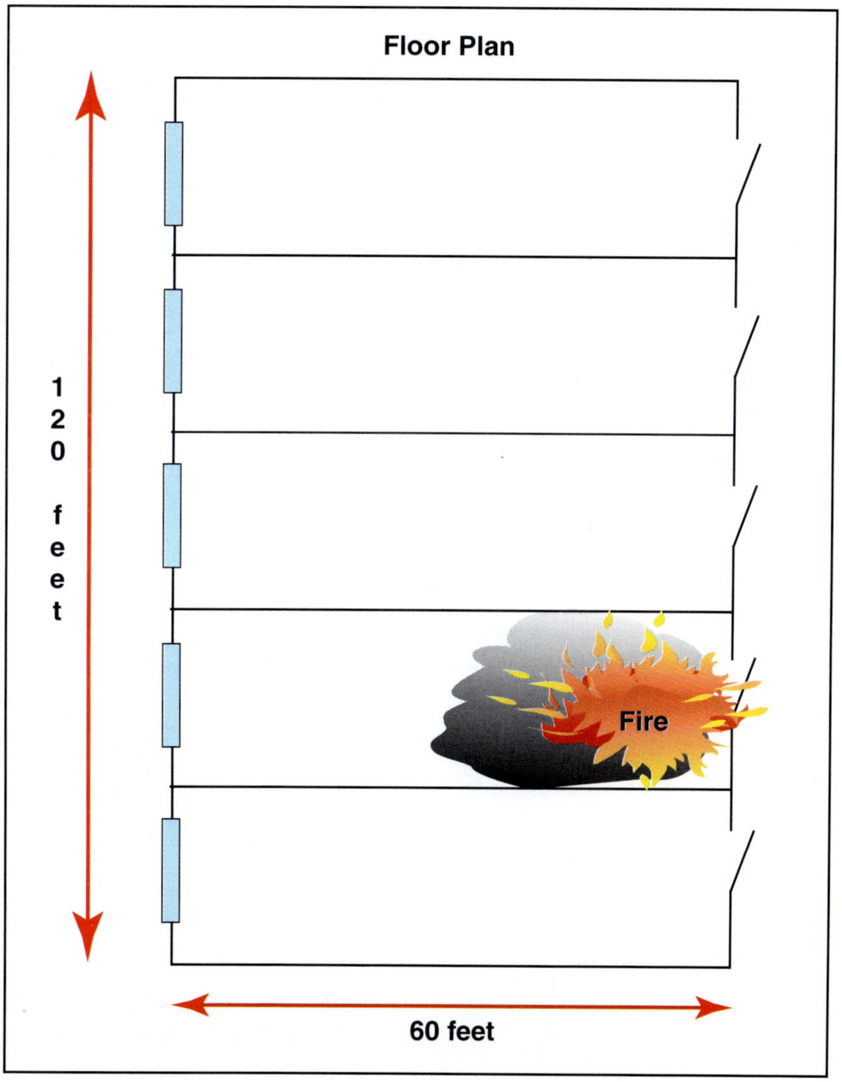

Activity 2.1 (cont'd)
Checklist

Critical Cues

Water 2,000 gpm available

Area

Life

Location/Extent

Apparatus/Personnel 2 engines, 1 truck, 1 chief

Construction

Exposures

Weather Clear, cool

Auxiliary Appliances

Special Matters

Height

Occupancy

Time 0245 hours, Monday

Problems Identified

1.
2.
3.
4.
5.
6.

Activity 2.2
Determining Strategy

Purpose
To build on the information developed in Activity 2.1 and to produce a list of applicable strategies for the incident.

Directions
1. Review the scenario and your answers to Activity 2.1. Determine the strategies that are applicable to the first-in officer. This means not concerning yourself with overhaul and salvage.
2. Write the applicable RECEO VS (Rescue, Exposures, Confinement, Extinguishment Overhaul, Ventilation, and Salvage) strategies below.
3. From the seven factors above, choose only those that initially affect the first-in officer.

   a. ______________________________

   b. ______________________________

   c. ______________________________

   d. ______________________________

   e. ______________________________

   f. ______________________________

Activity 2.3
Action Planning

Purpose

To build on the information developed in Activity 2.1 and Activity 2.2, and to produce appropriate tactics for the strategies selected in Activity 2.2.

Directions

1. Review the strategies that you determined in Activity 2.2
2. For each strategy, select the most appropriate tactics.

Strategy 1 ______________________________

Tactic ______________________________

Tactic ______________________________

Tactic ______________________________

Strategy 2 ______________________________

Tactic ______________________________

Tactic ______________________________

Tactic ______________________________

Strategy 3 ______________________________

Tactic ______________________________

Tactic ______________________________

Tactic ______________________________

Strategy 4 ______________________________

Tactic ______________________________

Tactic ______________________________

Tactic ______________________________

Strategy 5 ______________________________

Tactic ______________________________

Tactic ______________________________

Tactic ______________________________

Chapter 2
Unit Questions

1. What are the six steps in the classical decision making methodology?
2. What are the four steps in the recognition-primed decision making methodology?
3. The study by Klein and Associates found evidence of simultaneous comparisons or the relative evaluation of two or more options in fewer than what percent of incident commanders?
4. What is a cue, in fire service terms?
5. When is it appropriate to use the classical method of decision making?
6. How does one learn to use the recognition-primed decision-making method?
7. What are the three action steps and three results in the command sequence?
8. What is size-up?
9. What preincident information is available to fire departments?
10. What is considered to be the first objective of size-up?
11. List each element of WALLACE WAS HOT.
12. Why is it important for an incident commander to be able to make predictions?
13. What are Lloyd Layman's seven factors for the development of strategy?
14. What are tactics?
15. What elements comprise the action plan for an incident?

Suggested Activity Answers

Activity 2.1

Critical cues:

Water: 1,000 gpm available

Area: 7,200 sq. feet

Life: none apparent or probable

Location and extent: 50 percent involvement of one structure with probable extension into attic space

Apparatus and personnel: two engines, one truck, one chief

Construction: ordinary (masonry, wood-joist), concrete slab

Exposures: none outside of the strip shopping center

Weather: clear, cool

Auxiliary appliances: none

Special hazards: false mansard

Height: 12 feet

Occupancy: mixed mercantile

Time: 0245 hours

Problems identified:

1. 50 percent fire involvement in one occupancy
2. Smoke in the attic space
3. Probable fire spread to the common attic space
4. Smoke in adjoining occupancies

Activity 2.2

Strategies:

Rescue (primary search that is done at all incidents); confinement; extinguishment; ventilation

Activity 2.3

Strategy 1: Call additional resources to the scene

Tactic: Strike a second alarm of two engines and one truck.

Strategy 2: Rescue

Tactic: Engine 1 to quickly search adjoining occupancies to the right side of the fire.

Tactic: Engine 2 to quickly search adjoining occupancy to the left of the fire.

Strategy 3: Confinement

Tactic: Engine 1 to position a hand line in the occupancy to the right of the fire.

Tactic: Engine 2 to position a second hand line in the occupancy with Engine 1.

Tactic: Engines 3 and 4 to position hand lines in the occupancy to the left of the fire.

Strategy 4: Extinguishment

Tactic: Engine 1 pump operator to spot E-1 in front of fire occupancy and apply master stream device to the fire.

Strategy 5: Ventilation

Tactic: Truck 1 to vent the roof over the fire occupancy to stop horizontal extension

Tactic: Truck 2 to open the building horizontally front and rear.

Explanation:
The actions of Engines 1 and 2 prevent spread of the fire toward the larger portion of building (3 occupancies). These companies would pull ceiling and inject streams to stop the horizontal extension. Engines 3 and 4 prevent the spread of the fire into the single adjoining occupancy on the left. Truck 1 does the most critical job of opening the attic space to alter the air currents in the attic space and to draw the smoke and heat from the building. Truck 2 must open the building horizontally to provide a tenable atmosphere and visibility for operating forces in the building. Engine 1 pump operator, who spots the apparatus correctly in front of the fire occupancy, applies water with the deck gun to extinguish the fire.

Suggested Activity Answers

1. Gather information

   Analyze the information

   Determine the problems

   Prioritize the solutions

   Select the tactics

   Issue directives

2. Look for certain critical cues

   Relate cues to previous similar situations

   Recall previous conclusions, results and action that best fit the new situation

   Issue directives

3. 12 percent

4. A cue is a sensory observation, usually verbal or visual, which by itself or in conjunction with other cues, leads a decision maker to relate to previously encountered situations (experience or training).

5. When in training

   When evaluating and planning and when time is not a factor

   When working with case studies

   When at a real incident and the cues are not readily seen due to inexperience with such an incident type

6. It is an experience-based methodology. One must get all the experience possible, either through actual incident experience or through training.

7. Size-up; problem identification

   Strategy and tactics; action plan

   Implementation; tasks

8. The process of gathering and analyzing incident information that affects decision making.

9. Knowledge of buildings, water supply, construction, etc.

   Prefire planning documents

   Building inspection documents

   Environmental considerations

   Time of day, day of week, season of year

   Knowledge of surrounding area

10. The first objective of size-up is problem identification.

11. Water; Area; Life; Location/extent; Apparatus/personnel; Construction; Exposures; Weather ; Auxiliary appliances; Special hazards; Height; Occupancy; Time

12. It is critical to determining the problems at an incident. Not everything is visual or verbal. The IC must be able to predict situations based on experience.

13. Rescue; Exposures; Confinement; Extinguishment; Overhaul; Ventilation; Salvage

14. Tactics are the operations that must be completed successfully to accomplish the strategy.

15. Strategy—What

    Tactics—How

    Action Plan—Who, Where, and When

    Communications plan

    Medical plan

Chapter 3
Practical Application of ICS

Practical Application of ICS

Chapter 3

Introduction

The Incident Command System (ICS) was born out of the wildland fires that ravaged Southern California in the late 1960s and early 1970s. Prior to this time, methodologies for incident management were agency specific. Due to the size of these incidents, large numbers of resources were committed, but because of agency differences significant problems became apparent. These problems included:

- *Command Procedures.* There were no clear cut lines of command. Responsibilities for various command positions were not clearly defined. Personnel management and scene safety were afterthoughts.
- *Resource Management.* There was no standardization of equipment, training, or staffing levels. Lack of procedures for requesting, staging, and assigning resources led to freelancing and a less than efficient use of resources. Additionally, the lack of accountability systems placed personnel and equipment in jeopardy.
- *Communications.* Departments used different frequencies, different radio systems, and different terminology, which created a communications nightmare.

Obviously, a better system needed to be developed to combat these costly and complex incidents.

Representatives from federal, state, and local fire agencies got together to address this issue, forming an organization known as FIRESCOPE (**FI**refighting **RES**ources of **C**alifornia **O**rganized for **P**otential **E**mergencies). This organization's mission was to focus on a variety of problems associated with large, complex wildland incidents. The Incident Command System (ICS) was developed by FIRESCOPE, and its success at controlling wildland fires led to its application in structural fire fighting. Modified and refined over the years, ICS has evolved into an all-risk management system that can be utilized by any size department on any emergency or nonemergency situation requiring the management of resources.

The National Fire Academy has adopted FIRESCOPE ICS as the standard to be taught in all of its operational courses, with the goal of reducing the loss of life and property, enhancing firefighter safety, and safeguarding the environment.

The National Inter-Agency Incident Management System (NIIMS) was developed to provide a common system that fire protection agencies could utilize at local, state, and federal levels. More recently, NIIMS has been adopted not only by many police agencies, but also by other non-public-safety agencies that may be involved in disaster management. NIIMS consists of five major subsystems that collectively provide a total systems approach to risk management. These subsystems include:

- The Incident Command System (ICS), which includes operating requirements
- Standardized training, which supports the effective operation of NIIMS
- A national qualifications and certification system, which provides a standard of training, experience, and physical requirements in order to fill specific ICS positions
- Publications management, which includes development, publication, and distribution of NIIMS support materials
- Technological support such as ortho-mapping, infrared photography, and a multiagency coordination system, which supports NIIMS operations

During this same period, the Phoenix Fire Department developed another incident scene management system, the Fire Ground Command system (FGC). In addition to delineating management functions and organizational structure, the FGC system includes subsystems for determining strategy, tactics, mode of operation, and so forth. Many fire departments across the country adopted and applied this system with great success.

As a result of these two management systems' (ICS and FGC) inherent differences in focus and development, there were inconsistencies in terminology and application. A single incident management system is critical to effective command and control of major incidents. It is on these types of incidents that a single department will interface with other resource agencies in their local area as well as resource agencies from state and federal governments. It is obvious that a single command system is critical to reducing the inherent confusion of large-scale incidents.

In 1990, recognizing the conflict between the two systems, the Phoenix Fire Department and FIRESCOPE created the National Fire Service Incident Management System Consortium. The initial mission of the consortium was to develop a standard incident management system that would be the national model. This organization is comprised of individual fire service leaders as well as representatives from most major fire service organizations. After a significant amount of work by these dedicated professionals, the IMS Consortium completed its first document, *Model Procedures Guide For Structural Fire Fighting,* in 1993. The major changes include:

- Organizational structure for the new Incident Management System would be from FIRESCOPE ICS.
- *IMS Procedures Guide* contains the subsystems for determining strategy, tactics, mode of operation, and application of the system.
- Administration Section is changed to Finance/Administration.
- Sectors are recognized and included as equivalents to Divisions/Groups.
- Responder Rehab is formally recognized and placed under the Medical Unit in the Logistics Section.
- Finance/Administration has responsibility for legal issues relating to the incident.

While the primary focus of the consortium's work related to structural fire fighting, the need for other, more specialized model procedure guides was recognized. The consortium developed subcommittees to meet this challenge and began to develop model procedure guides addressing areas such as high-rise, hazardous materials, EMS, wildland-urban interface, and urban search and rescue. In order to maintain standardization, all of these manuals will use the *IMS Model Procedures Guide for Structural Fire Fighting* as "boilerplate."

The Need for a System

The majority of incident scene problems are not the result of poor tactical actions. Generally, firefighters are well trained and carry out tactical assignments with a great deal of efficiency. The problems occur at the command and control level. Unfortunately, some departments feel that emergency operations are so unique that conventional rules do not apply and new ones must be invented. Or worse, they refuse to budge from the procedures used many years ago. A decade ago, most departments did not have an organized incident management system and certainly did not practice ICS as we know it today. However, today many departments are developing formal incident management systems, complete with written procedures. More importantly they are using these systems on the incident scene. ICS is no longer a buzzword, it has become a byword, a principle tool in the Incident Commander's tool box.

Regardless of the nature of the incident, all emergency operations have one requirement : *all operations must have organization to be successful.*

A Systems Approach

One of the best ways to understand an organization's nature is to view it from a systems approach. Over the past twenty years many knowledgeable people have come to consider the systems approach to business management as the single most influential concept of contemporary organizational theory.

What exactly is a system? Definitions may vary depending upon whom you ask, but the dictionary defines a system as a group or unit of interrelated parts or functions designed to achieve a common goal. Visualize a fire truck with all its individual components working together so that the apparatus functions correctly (Figure 3.1). All of the body's components are important to its success as a total system, even though some components may be utilized more than others. Some components' failure will have a greater impact than others' on the total system. The maximum efficiency of any particular unit, however, is not as important as achieving optimum efficiency for the overall organization. The systems-

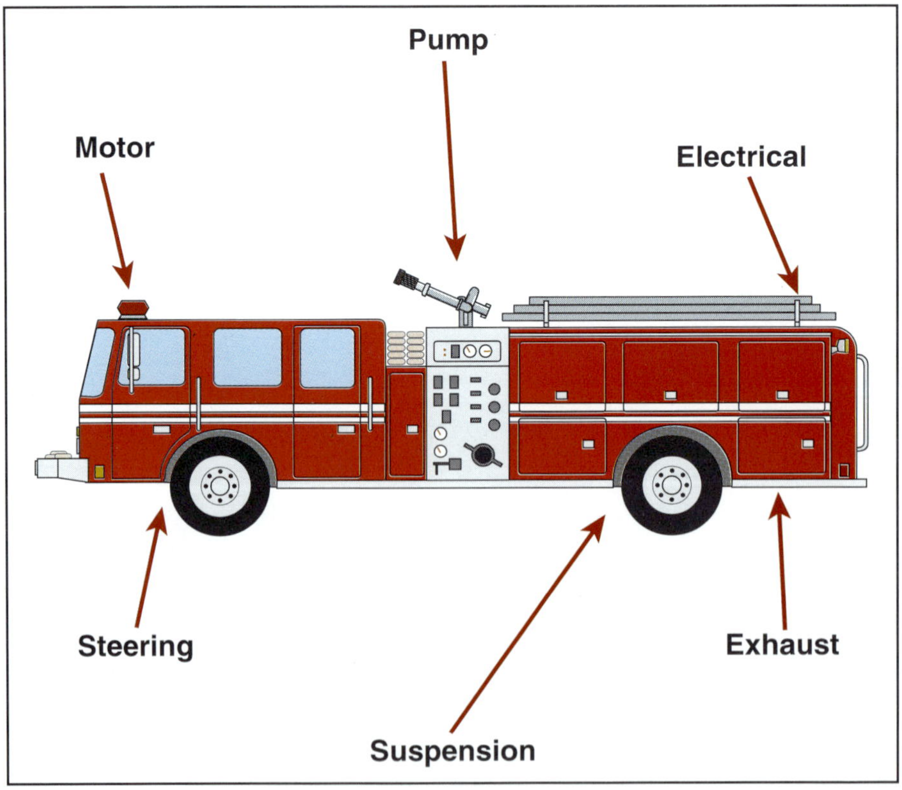

Figure 3.1. Fire truck systems. A system can be defined as several interrelated parts or functions designed to achieve a common goal.

oriented manager should focus on the whole organization and the optimum achievement of organizational goals.

The systems approach is not a new concept; there are many examples of its use. One is the automobile industry. The goal of an automobile manufacturer is to design and produce a vehicle that the public will like and purchase. To accomplish this, various units within the manufacturer's system must do their jobs—marketing, to research public appeal; engineering, to design the vehicle; assembly, to build and paint the vehicle, and so forth. Many of these units work independently at times, but all are important to achieving the goal. (Figure 3.2 shows typical corporate structure, with corresponding units of the IMS system listed in parentheses.) The military is another example of the systems approach. How could the combat soldier be victorious in battles without planning to evaluate the enemy and develop strategies; logistics to provide transportation, equipment and food; medical to care for the sick and wounded; and other units such as artillery and air support?

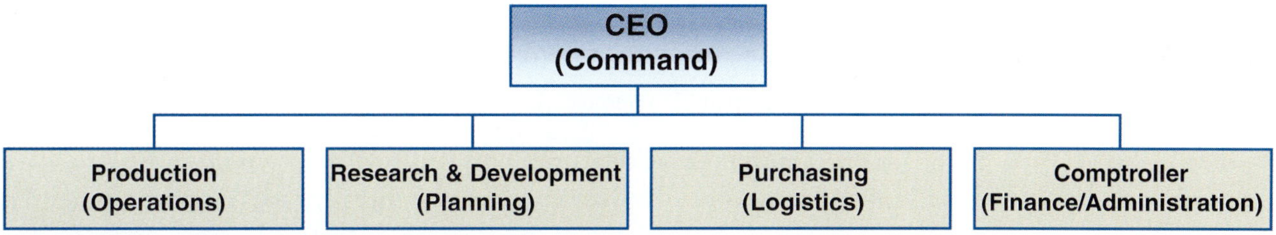

Figure 3.2. Typical corporate organization.

Why Organizations Use a Systems Approach

Organizations use the systems approach for many reasons. Once an organizational goal is established, all of the operations necessary to achieve that goal can be identified. These operations can then be grouped into related functional areas. By doing this, *functional responsibility* can be identified. In other words, the individuals responsible for each product or phase of the operation are designated. In addition, accountability can be established because individuals within a functional area can be given related responsibilities; by recognizing responsibilities and the position to which they are assigned, the organization will avoid or at least reduce the likelihood of duplication of effort.

The systems approach also provides a mechanism for *coordinating* the efforts of all areas involved by *delegating* authority and responsibility. Giving someone else the power to carry out responsibilities develops a clear line of authority in an organization and creates a manageable "span of control" (the number of individuals or jobs supervised by one individual). Delegating authority means that one individual does not have to oversee all facets of all operations, and it promotes the concept of "unity of command" (one person—one boss).

Identifying functional areas also facilitates identifying necessary tasks. Concentration on one area rather than on many tends to ensure that all tasks relative to that area are addressed. Equally important is the prioritizing of tasks. When objectives are limited within the scope of the functional area, tasks necessary for their accomplishment are more easily identified and performed in the proper order. In addition, people are less likely to freelance (develop and carry out their own objectives).

Safety is, or should be, a major concern of all organizations. In the last decade, a significant emphasis was placed on worker safety. Pressures from all levels of government have created a variety of laws and agencies dedicated to ensuring a safer working environment, both short- and long-term. Safety is an attitude. It is the responsibility of everyone in the organization, but it is of primary importance to the supervisor and/or manager. The systems approach allows those responsible for functional areas to monitor and enforce safety regulations for their areas. An operation that lacks organization is usually one in which various tasks and responsibilities are overlooked. Invariably, safety is one of the first areas to fall into that category. The systems approach also allows for the forming of a separate safety function in those organizations large enough or complex enough to warrant one.

Whenever operational problems are discussed, communications always finds its way into the conversation. Generally, too much nonessential information is routed to all areas, and too little relevant information is routed to the proper places. The systems approach can identify who should receive what information. The various functional areas can then identify the information that is relevant to their areas or objectives. Listening and comprehension are enhanced because time and effort are not wasted scanning vast amounts of information to determine whether or not it is pertinent.

Now let's apply what we have discussed so far to the day-to-day operations of a typical fire department. Most departments are comprised of two or more divisions responsible for specific areas, such as suppression, EMS, training, or fire prevention. This identifies the functional areas. Generally these specific areas are under the command of a chief officer or another appropriate authority. This delegates functional responsibility. Personnel assigned to a functional area answer to the superior responsible for that area, thus maintaining clear lines of authority and communications. Each area is responsible for its objectives according to the overall goals of the department, and generally, each area sets its own priorities for accomplishing those objectives. Company officers decide what gets done first in order to accomplish their assigned tasks.

Characteristics of the Incident Command System

Although any individual incident command system will be unique to its specific circumstances, all will share the following characteristics.

Organizational Structure. All management systems, if they are to be effective, must have an established organizational structure. This structure provides the hierarchy of responsibility, lines of authority, and formal communications. It further provides all those in the organization with a clear understanding of their own authority, power, and responsibility.

Common Operating Procedures. In emergency operations, we must be able to move rapidly from a nonincident organization to an incident organization. If we are to be effective, we must have a standard methodology that allows us to react out of habit and immediately transcend to the basic level of our emergency incident organization.

Common Terminology. If people are to communicate effectively, words must have a single definition, functions must have one set of responsibilities, and no two words may have the same definition. If we challenge this axiom, then we introduce confusion into information and orders, thus reducing the efficiency and safety of our personnel.

Personnel Qualifications. ICS is not a rank-oriented system. It is a performance-oriented system. This simply means that the best qualified person is placed into the appropriate function for any given situation. All personnel, regardless of rank, who are going to be involved *at any level* in the system, must be trained in the use of ICS.

Incident Commander. Under most circumstances, the only position (function) in the ICS that must always be staffed is that of the Incident Commander. The exception to this is that in the event of a hazardous materials incident the Safety Officer position also is required to be staffed. The point is that someone who is responsible and accountable must always be in charge. This IC position must be assumed by the first-ranking member of the fire department to arrive at the scene.

All-Risk System. While originally designed for wildland-urban interface fires, the ICS has evolved into an all-risk, all-situation emergency management system. We manage people, not fires, floods, earthquakes, tornadoes, hazardous materials, or mass casualties. ICS is a resource management tool, keeping in mind that our greatest resource is people. It has been tested on all types and sizes of emergency situations and nonemergency situations alike. Fire departments around the country have used it to manage trade shows, recruit academies, retirement dinners, holiday events, and VIP visits.

Jurisdictional Authority. Unless federal, state, or local law states otherwise, the ranking officer from the involved jurisdiction is the Incident Commander. The IC cannot be removed from this position unless the law dictates otherwise. The IC may, however, transfer command to a person from another agency or jurisdiction at his or her discretion.

Span of Control. Span of control is the number of personnel or functions reporting to any given individual. Generally the parameters range from two to seven, depending on the complexity of the situation. The ICS is based on a span of control of five. Certainly in any situation not yet under control, no one operating under ICS should have direct control over more than five personnel or functions.

Deputy and Assistant Positions. *Deputy* is the functional title for an individual who may provide relief for a superior and in the superior's absence could be delegated the authority to manage a functional responsibility or carry out a specific task. These individuals must be fully qualified and capable of fulfilling the assignment for which they are serving in the capacity of Deputy. For example a Deputy Incident Commander should be trained and qualified to perform as an Incident Commander should the need or opportunity arise. Deputy positions are most often utilized on very large incidents that span two or more operational periods and where fatigue may become a problem requiring relief. Another appropriate use for the Deputy position would be in a Unified Command situation when the primary responsibility may shift from one agency to another as the incident progresses and there is a need for shared control. For example, in a hostage or terrorism incident, the primary focus may shift from the first responder (usually the fire department) to the police and possibly back to either fire or medical after the situation is neutralized.

When utlitized, Deputies can be assigned to the Incident Commander, any of the General Staff positions, and Branch Directors.

Assistant is the title used for subordinates of various Command Staff positions such as *Assistant Safety Officer* or *Assistant Information Officer*. These individuals are subordinate to the primary position but should be knowledgeable and qualified to fill the primary position if necessary. On larger incidents, where a broader organization is implemented requiring more support activities, Assistants may be used to supervise Unit activities.

Unity of Command. Unity of command means that everyone has only one boss and knows who that boss is. When anyone receives orders from more than one person, confusion is the predictable outcome. The confusion in turn delays solving the problems at hand, increases life and property losses, and jeopardizes firefighter safety.

Everyday Application. If we are going to effectively manage large incidents when they occur, we must develop appropriate habits that will enable us to efficiently apply the ICS. The only way to develop a habit is by doing the same thing over and over. We must implement ICS on every incident to which we respond, using those functions that are appropriate to the complexity of the incident and available resources. It is also important to utilize ICS during training, just as if evolutions were real incidents. In short, the ICS must become second nature to us.

ICS Organization at Expanded Incidents

Usually, the ICS organization at routine incidents is simple. There are relatively few problems, a limited number of resources, and the IC can effectively handle all management functions. In many departments this is the case approximately 90 percent of the time. On a day-to-day basis, most responses can be handled with the first alarm assignment. As an incident increases in complexity, however, more problems develop, and more solutions are required. The increased numbers of resources result in increased management concerns. The IC becomes overwhelmed and is unable to do all of his or her jobs and provide all of the answers without assistance.

Attempting to deal with all management functions may overload the IC. When overload occurs, important details may be overlooked and personnel safety may be compromised. The solution is to use the Incident Command System to delegate specific responsibilities and authority to other individuals. Cues that should drive the Incident Commander to expand the organization include such developments as an increase in the incident's complexity, an increased number of resources, and additional management concerns. The IC should ask, "Can I handle everything?" If the answer is no, the organization needs to be expanded.

The Five Major Functions

The ICS organization is incident driven. As conditions or needs change, so must your organization. The modular design of ICS allows the organization to be structured to meet specific situations. The ICS provides a systems approach to effective incident management. It is important to remember, however, that the tool is only as good as the craftsman. To effectively employ the ICS, the Command Officer must understand its functional positions as well as each of those positions' roles and responsibilities.

The five major management functional areas of the incident command system are Command, Operations, Planning, Logistics, and Finance/ Administration (Figure 3.3). It is important to remember that the blocks

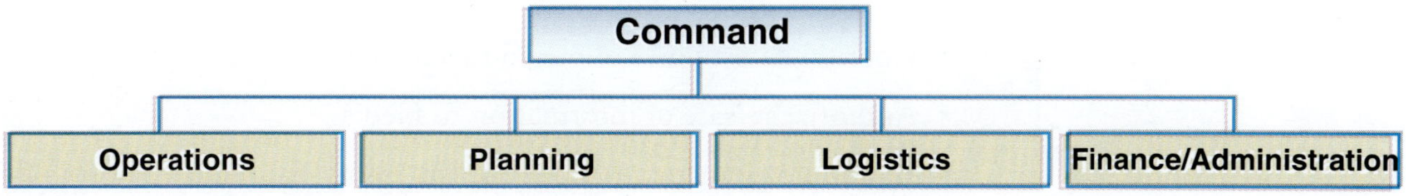

Figure 3.3. Functional areas of ICS.

in the ICS are "functions or jobs," *not* positions that must be staffed. It is also important to remember that as an Incident Commander, you remain responsible for any function or job that you do not assign to someone else. So, depending on the complexity of the incident and the numbers of resources at your disposal, you may find yourself "wearing several hats."

Command

The first component of the Model ICS is *Command.* No operation can be implemented unless someone is in charge. The first fire department member or company arriving on the scene should assume command and begin to take appropriate action. The initial IC will remain in command until properly relieved or until the incident is stabilized. In the initial stages of an incident, due to the limited number of personnel, the IC will normally carry out all of the Command Staff functions (Safety, Liaison, Information) and all of the General Staff functions (Operations, Planning, Logistics, Finance/Administration) until they can be assigned to qualified personnel. Overall the IC has complete authority and responsibility for the incident. If a higher-ranking officer wants to make a change in the management of an incident, that officer must first be on the scene and must then use the transfer of command procedure.

Command

Organizations should develop written procedures or guidelines for the use of ICS. This will not only fix the responsibility for command but will clarify other functions and agency-specific assignments within the incident management system. These procedures should address communications, reporting, transfer of command, mutual-aid interface, or any other significant issues that might effect the organization's implementation of ICS.

Responsibilities of Command. Command is responsible for overall management of the incident. These responsibilities include:

- Assess incident priorities and size-up the incident.
- Determine strategic goals.
- Determine tactical objectives.
- Develop an Incident Action Plan (IAP).
- Develop an appropriate organizational structure.
- Manage incident resources.
- Coordinate overall incident activities.
- Ensure firefighter and scene safety.
- Coordinate activities of outside agencies.
- Authorize release of information to media.

Clear communications are critical in implementing command responsibilities. The "Command" radio designation should be used along with a geographical identifier (for example, "Main Street Command"). Once an assignment is made, the individual no longer uses a rank or unit designator, but rather his or her functional title.

Command Options. At the majority of incidents, the first arriving company officer will be the initial Incident Commander. However, guidelines should state that the first arriving unit or member will assume command, regardless of whether an officer is present. The type of incident, conditions, and available resources will guide the officer's direct involvement and mode of operation.

Nothing-Showing Mode. Normally in these situations, the first arriving company will investigate while other assigned companies remain in Level 1 staging. Under most circumstances, the company officer will remain in command, using a portable radio. However, some circumstances would make passing command prudent—investigating on the upper floors of a high-rise or deep in a very large building, communications problems, for example. Departments should evaluate these types of situations carefully when developing guidelines.

Fast-Attack Mode. When stabilizing/mitigating the situation requires immediate action, the officer's direct involvement will probably increase the chances of success. In situations that involve problems like a known rescue or firefighter safety and immediate intervention is crucial to success, the company officer can command the incident using a portable radio. Certainly, staffing levels will dictate how long this situation might continue, but normally it will last only a few minutes. If the situation remains unresolved as more resources arrive, the company officer will either pass command or exit the situation and establish a command post.

Command Mode. Because of their complexity, size, or threat to the community, certain incidents require immediate and firm command. In these cases the company officer should establish a formal command post and assume command until relieved by a ranking officer. When a company officer removes himself or herself from the company to assume command, the crew must not be left to freelance. Any of several options might be appropriate, depending on staffing levels and available resources. The remainder of the crew could be assigned to command post functions or as a Rapid Intervention Crew. They could be combined with another company and assigned to that officer; or at the very least, one of the crew should be placed in charge and be provided with a portable radio.

Passing Command. Passing Command is generally associated with a critical fast attack situation or known rescue. In these situations the first arriving officer, recognizing he or she will have to become tactically involved, passes command to the next arriving officer. Generally speaking, tactical involvement reduces the officer's effectiveness as an Incident Commander. In many departments, due to limited staffing or extended response times, however, it may be the only available option. If this is an option, passing command may be advantageous and benefit the operation in some situations. Passing command alerts the other unit leader to change from an "order-taking" mode to an "order-giving" mode. The officer assuming command must be on scene, must declare so via radio, and must contact the first-due officer who passed command. This will prevent a gap in the command function that might create confusion and interrupt the continuity of command.

In addition, command should be passed only once (except under extraordinary conditions). Otherwise, "free-enterprise fire fighting" may result as command is passed from one unit to the next based on their sequence of arrival. It is imperative that this and all other parameters of passing/transferring command be clearly stated in written department guidelines and that all personnel be familiar with those guidelines.

Transfer of Command. Normally, command of an incident is transferred to improve the quality of the command structure and increase the probability of successful mitigation. While the principles of transfer of command remain constant, the actual application of the process may vary from agency to agency. The process can be affected by several factors including, but not limited to:

- Staffing levels
- Organizational structure
- Availability of personnel (full-time vs. all-volunteer personnel)
- Organizational policy
- Geography
- Response procedures

Whether an agency is affected by all, some, or none of these parameters, it is critical that the agency predetermine its transfer of command procedures and put them in writing.

A higher-ranking officer who arrives on scene is not necessarily prepared to assume command of the incident. This officer may or may not know of previous orders or grasp the current situation. Without a thorough briefing of the situation status, the newly arrived officer may compromise incident operations. This is why it is essential that a standard operating guideline (SOG) for the transfer of command be developed and practiced within the organization. It is also important to remember that command is transferred in *both* directions, up as the incident escalates and down during the demobilization phase.

The best method of transferring command is through a face-to-face meeting between the initial Incident Commander and the new Incident Commander. In face-to-face conversation, the relieving Incident Commander can take full advantage of all communication mediums. Remember, communications is more that just words; vocal pitch, facial expressions, hand gestures, and other body language assist greatly in conveying necessary information. The officer being relieved can also "read" the receiver's body language, helping to determine whether or not the officer assuming command understands the message.

The person being relieved should review the tactical worksheet with the officer assuming command. This sheet provides the most effective framework for transfer of command because, if properly used, it outlines the location and status of personnel and resources, providing an overview of current strategy and tactics as well as personnel accountability. The person being relieved should then be reassigned to the best advantage of the officer assuming command (Planning Chief, Safety, Division/Group/Sector Supervisor, etc.). Remember the relieving Incident Commander is at a disadvantage. The new Incident Commander has not been on the scene long and, with some actions already in progress, is in a catch-up mode. The information this officer receives and retains is critical to his or her knowledge of the situation and the success of the next operational phase.

The second best method of transferring command is by radio. However, because radios allow only spoken communication, radio transfer often leaves the relieving Incident Commander with information gaps and extends the time needed to catch up to the incident. These information gaps can lead to poor initial decisions and affect firefighter safety. In addition, by using the radio to transfer command, you tie up the frequency. Depending on the intensity of the incident and the number of channels available, this could hinder operations.

The least desirable method of transferring command is to do so without an information exchange. This method should be used only as a last resort. The new Incident Commander is usually at such an informational disadvantage that catch-up time is significantly extended.

Again, it is critical that a briefing takes place when command is transferred. This briefing should include, as a minimum, the following information:

- Present incident status/conditions (rescue situations, injuries, hazards, etc.)
- An Incident Action Plan (strategies/tactics being employed, operational mode)
- Progress toward achieving incident objectives
- Safety considerations and concerns, including a Personnel Accountability Roll call (PAR)

- Assignment/deployment of companies and personnel operating on the incident scene
- Projection of incident conditions and additional resource needs, and appropriate recommendations

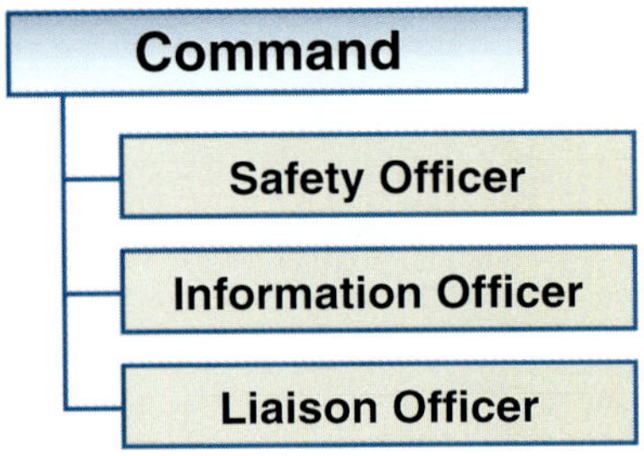

Figure 3.4. Command staff.

Command Staff. While Command is responsible for overall incident activities, certain functions must be accounted for immediately. These functions cannot be delegated to other major functional areas and are considered *Command Staff* functions, reporting directly to the Incident Commander (Figure 3.4). They include the Safety Officer, the Information Officer, and the Liaison Officer.

Safety. Normally, the *Safety Officer* position is staffed when the Incident Commander cannot devote sufficient time to safety analysis. This is generally the case on all but the smallest of incidents. As the number of resources and the incident's complexity increase, it becomes imperative that the IC designate someone else to focus on personnel and scene safety.

The safety officer is responsible for enforcing established safety procedures and practices at the incident. In addition, the Safety Officer should analyze the emergency scene, identifying present and potential hazards or unsafe conditions. The Safety Officer should be mobile, moving about the emergency scene as needed to gain the best view of the incident's dynamics. In the case of complex or geographically large incidents, the incident Safety Officer may have one or more Assistant Safety Officers (ASO) assigned to various segments of the incident in order to provide a more complete picture of scene safety.

A Safety Officer has the authority to stop unsafe acts when time or conditions prevent using normal lines of authority. Safety must immediately notify the Incident Commander of such actions. Because Safety Officers are invested with such great responsibility and authority, they cannot be just anyone on the incident scene. For example, a Safety Officer for structural fire fighting must be very knowledgeable in fire behavior, building construction and collapse potential, and department safety rules and regulations and must possess considerable experience in incident strategy and tactics. In hazardous materials responses, the law not only mandates that a Safety Officer be appointed but also that the individual must be trained to the Hazardous Materials Technician level.

Many larger departments have full-time Safety Officers or, in some cases, even safety battalions. Most departments use personnel with other full-time duties such as fire prevention or training to also fill the Safety Officer's. How a department fills the Safety Officer's position is not important, as long as the department uses a properly trained individual.

Information. The *Information Officer* is responsible for gathering and releasing incident information to the media and other appropriate agencies.

This is an important position and requires training in effective communications. All media releases must be approved by the IC. These releases may be covered in agency specific policies that state the parameters within which the Information Officer may operate, or they may be standardized by a media release form that specifically outlines the information to be released.

The Information Officer has the responsibility to update Command on media needs, new information, and deviations from policy. Equally important, Command must keep the Information Officer current on the status of incident operations.

The information position is typically staffed when the Incident Commander cannot take the time to speak with the media. At that time, a media area should be established away from the Command Post and will serve as a point of contact for media representatives.

Liaison. The *Liaison Officer* is the point of contact with outside agencies and is responsible for identifying and coordinating these agencies. The Liaison Officer is underused yet can be very valuable, especially on large, complex incidents. This position is normally staffed when the Incident Commander can no longer devote sufficient time to converse with outside agency representatives. Individuals who are well versed in ICS and thoroughly trained in the liaison role can provide an invaluable service to the IC. They can act as a buffer between the IC and other agencies, determining who should go to the Command Post and who should remain in Liaison. As a rule of thumb, those agencies that have legal responsibility and/or provide significant human resources to incident operations should be at the Command Post. Those agencies that provide assistance but have no legal responsibility should remain with Liaison until assigned elsewhere.

A Liaison Area should be established near, but not at, the command post. It should be close enough for the Liaison Officer to communicate effectively with the IC but distant enough to reduce the noise and confusion around the Command Post.

General Staff. The *General Staff* comprises those people who perform the four major functions of Operations, Planning, Logistics, and Finance/Administration. Each of the General Staff positions may have one or more deputy positions if necessary. Deputies should always be as qualified as the person for whom they are working. This may be a very flexible position with the Deputy assuming a primary position, a relief position, or a specifically assigned task. The Deputy will often be a qualified individual from your own agency; however, in large emergencies, particularly those that involve multiple agencies or jurisdictions, it often enhances interagency coordination to assign a person from an outside agency to the Deputy's position.

At the section level, the person in charge will be designated as a chief. For example, the person in charge of the Planning Section will always be called the Planning Section Chief.

Few of us regularly see a fully developed Incident Command System, but we must remain aware of a number of organizational elements that can be activated as necessary. This is also true of the General Staff positions. Each of the major sections can expand internally to meet the needs of the incident.

Operations Section

The Incident Commander will determine the need to establish an Operations Section at an incident or event. Until Operations is established, the IC will have direct control of tactical resources. However, once the Operations Section is staffed, the IC will no longer talk directly with tactical units. The point at which Operations is established may vary greatly from agency to agency. Some departments will have specific parameters for establishing the Operations Section, while others may leave it to the discretion of the Incident Commander. Generally, Operations will be staffed on larger incidents involving, or projected to involve, 15 or more fire companies. However, the complexity of an incident may also drive the staffing of the Operations Section. For example, a hazmat incident might involve only a few fire companies but would still require the response of several outside agencies or technical specialists, increasing the IC's administrative duties and making it difficult to devote adequate attention to the tactical needs of the incident.

Once Operations is established, the section chief has responsibility for all tactical operations in all phases of the incident. This individual will develop and manage the Operations Section to meet the strategic goals developed by Command.

An incident should have only one Operations Section Chief. He or she usually (but not necessarily) will be from the jurisdiction or agency that has the greatest involvement in terms of resources or legal responsibility.

Responsibilities of Operations. Operations is responsible for managing all *tactical* operations at an incident. These duties include:

- Managing all resources directly engaged in incident operations
- Determining and directing tactical operations
- Allocating and assigning resources to accomplish control of the incident
- Participating in the development of the Incident Action Plan (IAP)
- Reporting directly to the Incident Commander

Operations Section Expanded. The Operations Section comprises several elements that may be implemented to enhance span of control and incident management. The ICS charts and solutions depicted in this text are correct, but it is important to remember that they represent *just one* and *not the only way* of managing the incident using ICS.

For the purposes of this text, the term *engine* will mean a pumping apparatus and the term *truck* will mean a power ladder, tower, or snorkel apparatus as described in NFPA specifications. The terms *engine* and *truck* also refer to the personnel at the scene doing specific tasks. While many departments do not actually have a truck company, the fact remains that "truck work" (ventilation, forcible entry, search, etc.) must still be accomplished.

Single Resources and Crews. A single resource is an individual company (i.e., engine, truck, ambulance, rescue squad, or hazmat unit). Personnel other than those who arrive at the incident on an apparatus are formed into working units called *crews,* with each supervised by a crew leader. Crew size may vary, but should conform to span-of-control guidelines. Crews should be provided with radio communications and radio call signs. They may be designated by color, by the crew leader's name or function, or by some other distinction (i.e., Red Crew, Crew Burns, Vent Crew, etc.). Crews are considered as a single resource and used in that fashion.

Task Forces and Strike Teams. A *task force* is a group of resources with common communications and a leader. It may be either preestablished and dispatched to an incident or formed at the incident for a specific task. Generally, task forces comprise from two to five single resources, but may include any combination of resources necessary to meet the immediate need. A task force also may include non-fire-fighting units; for example, in a civil disturbance situation a task force may include a police unit for protection (Figure 3.5).

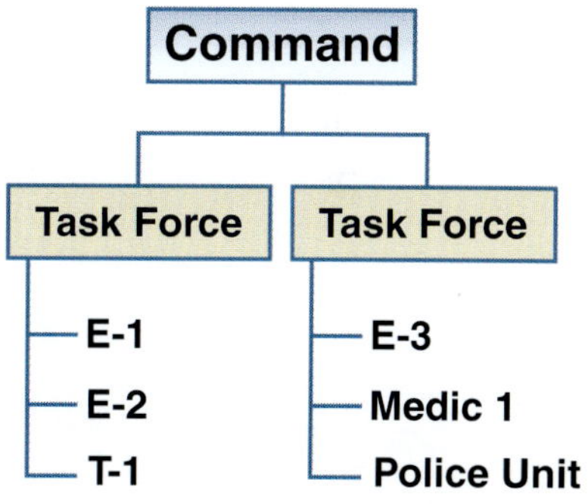

Figure 3.5. Command/task force organization.

The definition of a *strike team* is somewhat more specific. A strike team is a group of *five* of the *same type and kind* of resource sharing common communications and a single leader. While strike teams typically comprise engine companies, they also may comprise, for example, five Rescue Units, five Brush Engines, five Medical Units, etc. The strike -team concept is most frequently used at wildland-urban interface fires, but it certainly may be applied appropriately at other types of incidents or events. The exceptions to the rule of five resources comprising a strike team are found in wildland application. Hand crew strike teams are comprised of 30 personnel including a leader. A dozer strike team is comprised of two dozers and a tender. Division and Group Supervisors may use Strike Teams and Task Forces to correct span-of-control problems. Figure 3.6 illustrates how an IC's span of control over fifteen operating engine companies can be reduced to only three.

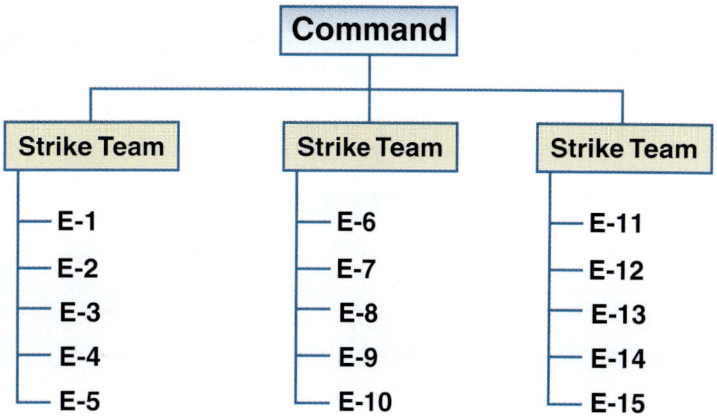

Figure 3.6. Command/strike team organization.

Let's review the use of single resources with an example. A first alarm assignment of two engines, one truck, and a chief officer, is responded to a reported fire in a grocery store (Figure 3.7). The first arriving company sees a typical grocery store of ordinary construction with 25 percent involvement and begins making assignments (Figure 3.8). Engine 1 has been assigned to advance a hose line into the building and begin to attack the fire. Engine 2 has been assigned to provide a backup line and begin a primary search. Truck 1 has been assigned to ventilation. The chief is Command.

How does this meet the definition of ICS? When the person in charge is called by radio, that person is not "Bob," "Chief 1," "Car 101," or anything other than "Command" or "IC." When you assume a position in ICS, that position's functional title becomes your radio designator. In most jurisdictions, over 90 percent of the responses can be handled with this assignment. This, then, is the "complicated" ICS you have heard about!

Staging. *Staging* is directly responsible to Operations. If Operations is not staffed, Staging will report directly to Command (Figure 3.9). The *Staging Area Manager* is responsible for the coordination, support, and distribution of incoming resources. Additionally, Staging must keep Command/Operations apprised of resources available in the staging area. Normally, the IC is the only person authorized to call additional resources to the scene.

Figure 3.7. Grocery store on arrival.

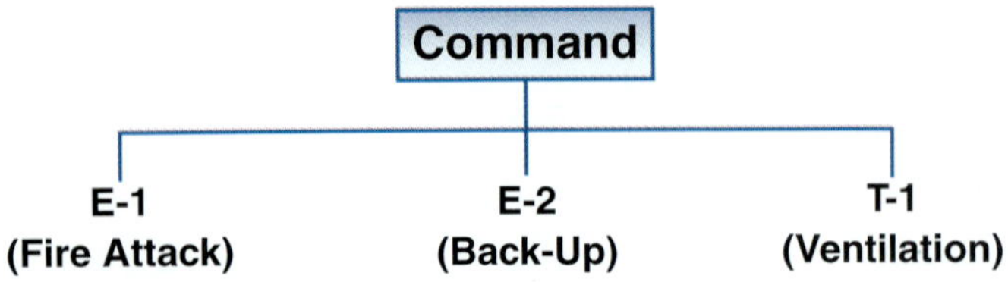

Figure 3.8. Command/fire attack organization.

Figure 3.9. Command/staging organization.

However, some departments have developed a written policy that stipulates minimum numbers of resources to be available in Staging at all times and authorizes the Staging Area Manager to maintain those minimums. For example, a policy might state, "A minimum of two engines and a truck will be in staging at all times." The Staging Area Manager would then be responsible to maintain those minimums and keep the IC informed when additional resources were called or when resources necessary to maintain those minimums were unavailable.

The Staging Area is a location where resources ready for immediate assignment (within three minutes or less) into the incident are temporarily placed. When designating a Staging Area, adequate space and proximity to the incident scene should be considered. Sites such as parking lots, schoolyards, parks, and shopping centers are ideal. Information on the Staging Area's location and manager should be broadcast when additional alarms are struck. For example, "Dispatch from Command, dispatch a second and third alarm; Staging will be located one block west in the Safeway parking lot; the Captain on the first-due second-alarm engine will be Staging Area Manager."

Level 1 Staging. There are two levels of staging. Level 1 staging is used to control the first alarm or initially dispatched units.

Have you ever responded to a structure fire that turned out to be a pot of food on the stove? Your dispatch could have been from three to six units, and they all might have parked right in front of the dwelling. Personnel from those units might have been tracking through the house to see what was going on. In addition to being unprofessional, such responses create an unnecessary mess. Or, maybe your department has the problem of companies assigning themselves on arrival (freelancing). Perhaps you are the Incident Commander at a working fire, and all of the responding companies are calling you for assignments. Even though you are not yet ready to make assignments, they keep calling you every twenty seconds until you give them an assignment, or else they arrive and immediately begin to freelance.

Ensuring that the department's standard operating guidelines (SOGs) include procedures for Level 1 staging can eliminate or greatly reduce all of these problems. Level 1 staging procedures should include these guidelines:

1. Allow only one or two units to go directly to the scene in addition to chief officers.
2. All other units shall stop approximately one block from the scene in the direction of travel.
3. Units shall report unit number and direction (for example, "Engine 2, 1 block west").
4. Units shall not proceed to the scene until directed by Command.

When developing procedures, remember that one block is an arbitrary distance; it is more important that apparatus in Level 1 Staging are positioned to maneuver safely (trucks may require greater distance than engines) and not in the immediate vicinity of the incident scene.

Communications procedures should also be included in any agency-specific policy. For example, the policy might stipulate that if a company is not assigned within two minutes after marking on scene at a working fire, it should repeat the message. It might further stipulate that after an additional two minutes the Company Officer should walk to the Command Post for a "face-to-face." In the initial confusion of a working incident, it is easy enough to lose track of a unit awaiting assignment. A process to remind, not badger, the IC is appropriate.

Level 2 Staging. Level 2 Staging is a location to which all greater alarm or mutual-aid companies, as well as personnel in private vehicles, report. The staging location should be announced when the additional resources are requested. For example, at the grocery store fire discussed earlier, the IC calls for two additional alarms, bringing four more engines and two more trucks to the scene. How will these resources be distributed as they arrive on scene? All could be responded directly to Staging, or some could receive assignments while en route. Those not receiving assignments would respond automatically to Staging once it is established (Figure 3.10).

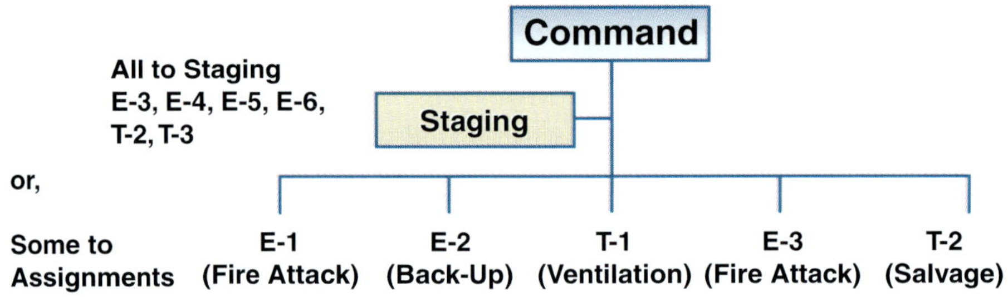

Figure 3.10. Staging organization.

Staging should not be thought of simply as a place to park apparatus. Staging at a high-rise fire could be on the tenth floor. Do not confuse *Base* with *Staging*. Base is used in conjunction with Staging at wildland, hazmat, high-rise, and other large, complex incidents. Each incident will have only one Base, while it may have more than one Staging Area. Base is a marshaling area for incoming resources at large incidents. They are organized at Base and then sent to Staging. This is where primary logistical functions are coordinated and administrated. Personnel are rested, showered, fed, and receive medical treatment at Base. Stockpiles of equipment projected to meet the long term needs of the incident will be located at Base. At a high-rise incident, Base is located at least 200 feet from the building. All companies report into Base and are then sent to a Staging Area two floors below the fire or directly to other assignments.

Sectoring System. To maintain span of control, track responsibility, and effectively manage the incident, a department must use a *sectoring system.* In the Incident Command System the "sectoring," or dividing up, of an incident scene is accomplished through the use of Divisions, Groups, Sectors, and Branches. FIRESCOPE ICS, as taught at the National Fire Academy uses only Divisions and Groups; however, many fire departments use Sectors effectively and efficiently. Which particular term might be used is not as critical as understanding each term's meaning, responsibilities, and position in the ICS organization. While Divisions, Groups, and Sectors may have different functions and responsibilities, they are all on the same level of authority and position in the ICS organization (Figure 3.11). Divisions, Groups, and Sectors work *with* each other and sometimes *across* each other's boundaries, but they *never work for* each other.

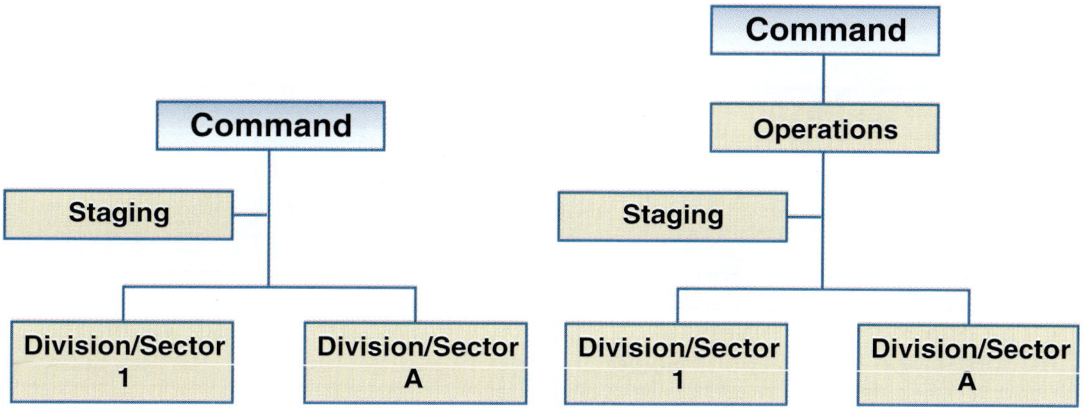

Figure 3.11. Command/staging/divisions organization.

Divisions/Sectors. A Division or Sector is an organizational unit responsible for resources and operations within a defined *geographical* area. This area might be a floor of a building, a side of a fire building, or a section of a brush fire. The Division/Sector is the level most often employed during routine fire department operations. When Divisions/Sectors are established on a horizontal plane, they are lettered (A, B, C, D, etc.) starting clockwise from a focal point, often the address side of the building (Figure 3.12). When divisions/sectors are established vertically, they are numbered by floor (Division 1/

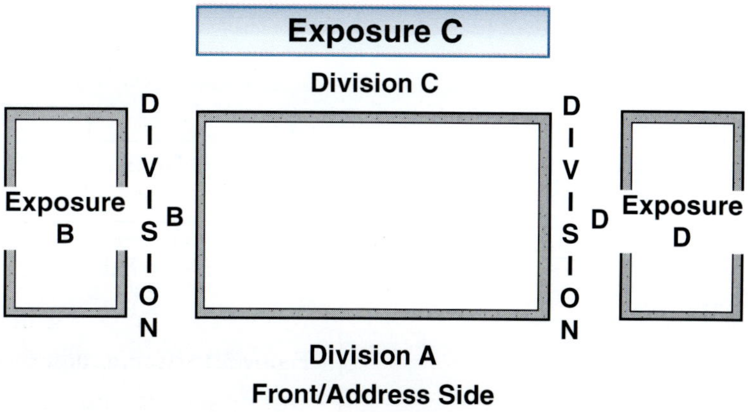

Figure 3.12. Horizontal division designation.

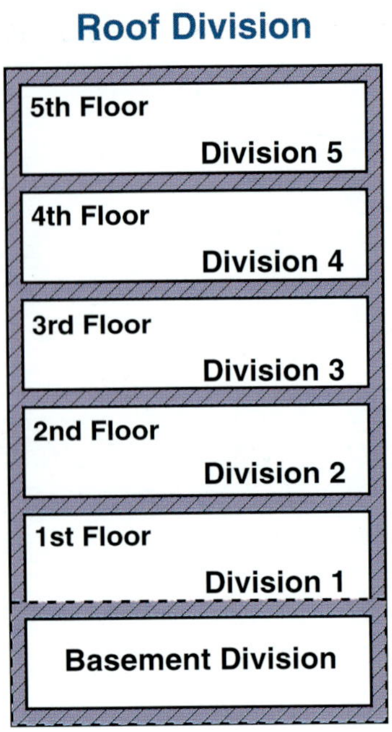

Figure 3.13. Vertical division designation.

Sector 1, Division 2/Sector 2, etc.; Figure 3.13). Groups/Sectors are resources assigned to a specific *function,* such as ventilation, rescue, or water supply (Figure 3.14).

Groups/Sectors are responsible for an entire job wherever it may be required. For example, a Ventilation Group/Sector would have ventilation responsibilities for the entire building. This being the case, Groups/Sectors will often work across Division lines and *must* coordinate with the Division Supervisors when they enter or perform work that will affect Division personnel, operations, or safety. Groups/Sectors should do nothing in a Division without coordinating with the Division Supervisor.

Application of Divisions and Groups/Sectors. Figure 3.15 diagrams a better way of using ICS to set up the grocery store fire discussed earlier in the chapter. This demonstrates the concept of establishing a habit of operation that will set the organizational foundation when you encounter a more serious situation.

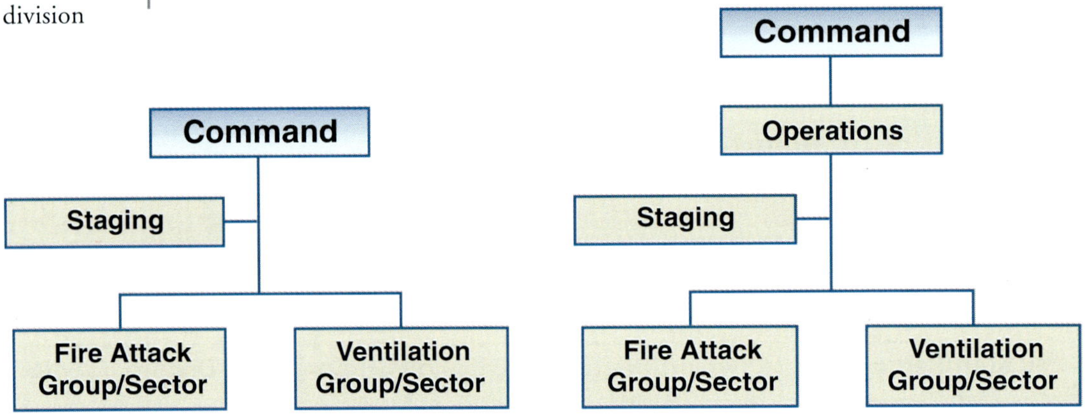

Figure 3.14. Command/operations/group-sector organization.

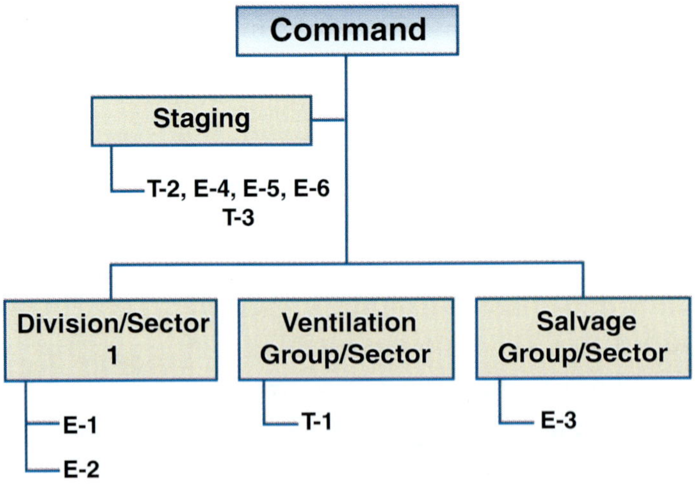

Figure 3.14. Command/operations/group-sector organization.

Example 1. You are in command of a fire in a three-story garden apartment (Figure 3.16). This apartment building is divided into three sections, each separated by a firewall. It is wood-frame construction with a ridge-pole-and-rafter roof assembly. The center section is well involved in the attic and third-floor levels. Figure 3.17 illustrates one way of organizing this incident:

Engine 1 arrives. The officer gives a brief initial report and directs the E-1 crew to attack the fire with a 1¾-inch hose line from the third floor and begin a primary search on that floor. The officer assumes Command. Command requests a second and third alarm.

Engine 2 arrives. Command directs the officer of E-2 to assume Division 3 and assigns E-1's crew to Division 3. Division 3 arrives and requests one truck company.

Figure 3.16. Three-story garden apartment fire on arrival.

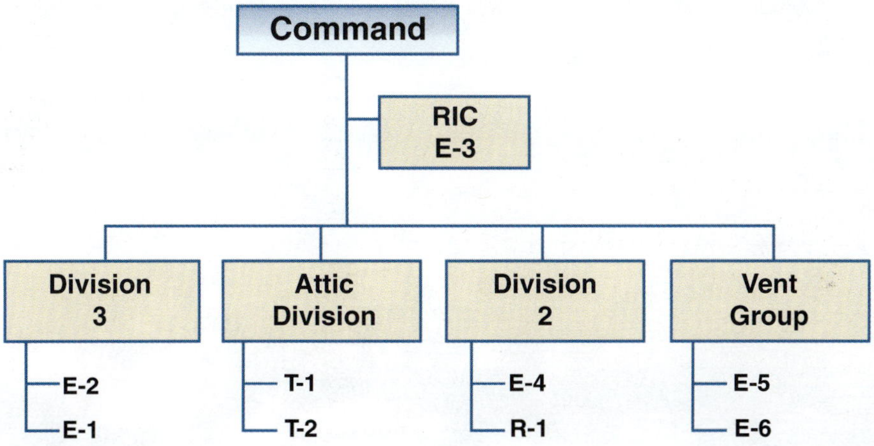

Figure 3.17. Incident command organization for three-story garden apartment fire.

Truck 1 arrives. Command directs T-1 officer to assume Attic Division and prevent the fire from spreading past the firewalls on Sides B and D. Attic Division requests one additional engine and one additional truck.

Battalion Chief 1 arrives. BC-1 assumes Command and retains the officer of E-1 to assist at the Command Post.

Engine 3 arrives. Command directs E-3 to become the Rapid Intervention Crew.

Engine 4 arrives. Command directs E-4 to assume Division 2 and to do a primary search and check for downward extension.

Truck 2 arrives. Command directs T-2 to report to Attic Division.

Engine 5 arrives. Command directs E-5 officer to assume Vent Group and to do horizontal ventilation where needed. Vent Group requests one additional company.

Rescue 1 arrives. Command directs R-1 to report to Division 2.

Engine 6 arrives. Command directs E-6 to report to Vent Group.

Example 2. Now let's look at a major fire in a four-story garden apartment (Figure 3.18). The attic and fourth floor are well involved. The IC at this incident faces complexities beyond those in the previous example. This is a nighttime fire in a larger building. It is February in a northern state, and the fire has displaced 48 families. The water system into the complex is a dead-end main, resulting in poor water supply. The only access road into the apartment complex provides no vehicle access to the rear (Side C).

Figure 3.18. Problems are increased in a nighttime fire.

Command is faced with many problems:

- Planning
- Relocation and care of displaced residents
- Water supply issues
- Limited access to the complex
- Nighttime fire
- No access to rear
- Logistics for fire and EMS units
- Liaison with other agencies

Recognizing the possibility of becoming overwhelmed, Command chooses to staff the Operations Section Chief function (Figure 3.19).

Engine 1 arrives. The officer gives a brief initial report and directs E-1's crew to attack the fire with a 1¾-inch hose line from the fourth floor and to begin a primary search on that floor. The E-1 officer assumes Command. Command requests a second and third alarm.

As the first and second alarm resources begin to arrive, Command initiates Division 4, Division 3, Vent Group, and Exposure Group.

Battalion Chief 1 arrives. Command is transferred and BC-1 assigns the E-1 officer to Operations. With the need to deal with displaced families, poor water supply, access problems, and other agencies, Command cannot continue to be involved with tactics. Command must step back and manage the overall incident, allowing Operations to handle the tactical issues.

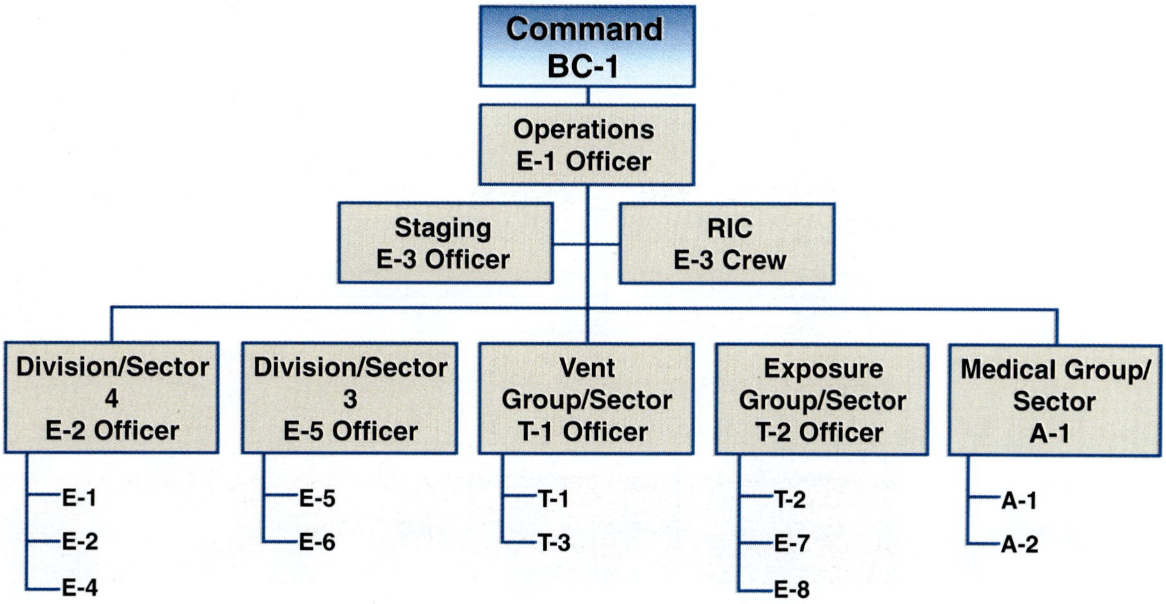

Figure 3.19. Fully developed incident command organization.

The E-1 officer could have been assigned to Planning or Logistics, but Operations seemed to be the area of greatest need, considering available resources. Another Chief Officer arriving on scene could have been assigned as Operations.

Are there other ways to successfully handle the incidents in these two examples? Certainly. Our suggestions are not the only ways to handle the incidents using ICS. How your department handles similar incidents will depend on many variables, including, among others, available resources, staffing levels, agency specific policies, training levels, and mutual aid agreements.

Branches. Let's return to the four-story garden apartment fire. The Operations Section Chief has established five divisions or groups. Due to the changing situation, the Operations Chief is now going to create a Division 2 and a Salvage Group (Figure 3.20). What has happened to the Operations Chief? The span of control has been exceeded. We can use branches to correct this span-of-control problem (Figure 3.21).

Depending on the magnitude and/or type of incident, Operations may break the incident areas into smaller segments for management purposes. These segments are referred to as Branches. Wildland firefighters often use a Branch for ground operations and another for air operations when responding to large brush fires. While few fire departments have the capacity for air operations, the concept is sound. An emergency requiring fire fighting and a large-scale EMS operation would be conducive to separate Branches. Large-scale disasters such as airplane crashes, terrorist incidents, floods, earthquakes, and other natural disasters are excellent candidates for this level of organization.

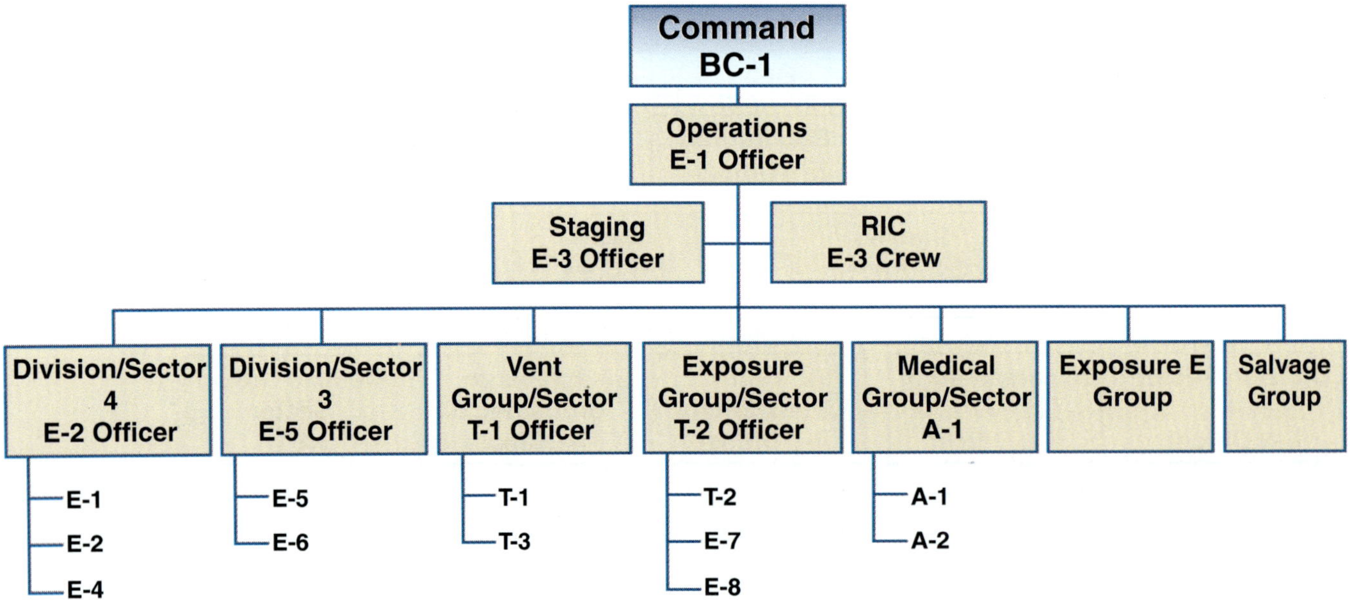

Figure 3.20. Span of control for operations is over five to one

Branches are the Incident Commander's or Operations Chief's tool for correcting a span-of-control problem created by too many Divisions/ Groups/Sectors.

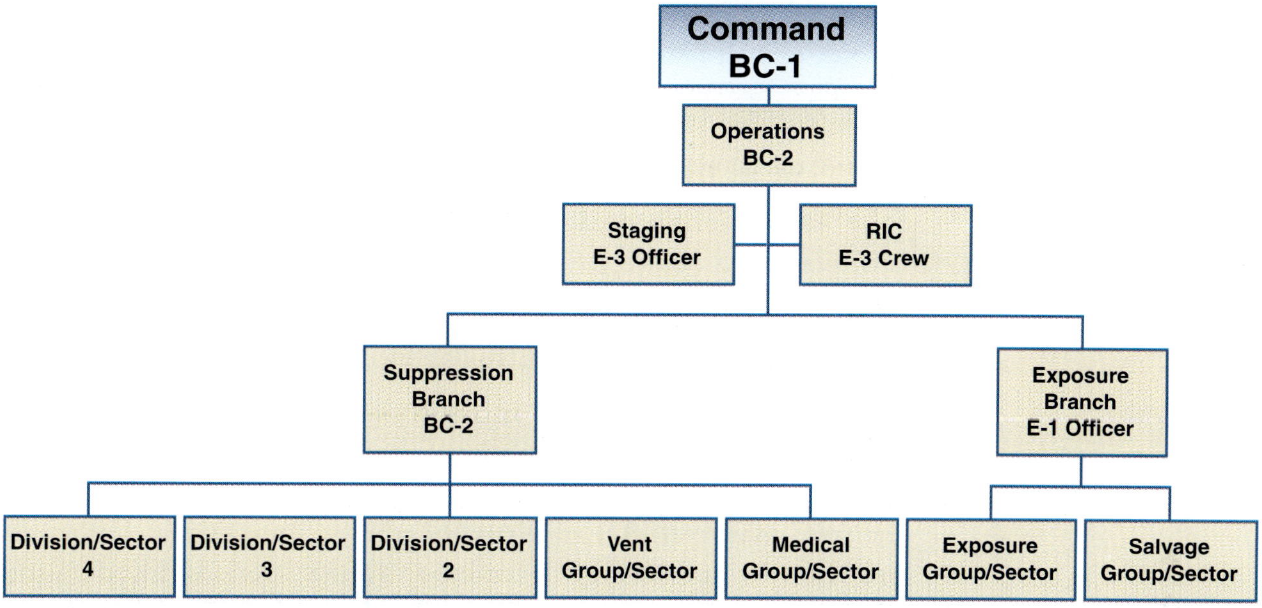

Figure 3.21. ICS Branches organization.

However, when an incident is already complex by the time the first units arrive and probably will require a large ICS organization, the initial IC should start at the Branch level instead of at the Division and Group level (Figure 3.22).

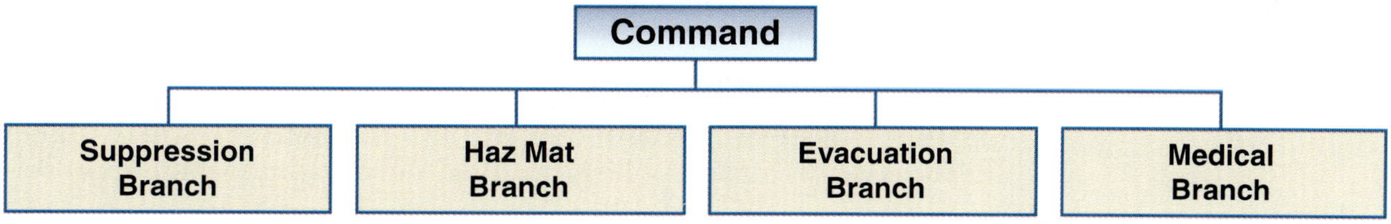

Figure 3.22. Command/branch level organization.

Planning Section

The Planning Section can be a vital element in incident mitigation. This function can be compared with the research and development section of an industrial corporation. The Planning Section gathers information, analyzes data, develops projections, and provides alternative solutions. This position often is either not staffed soon enough or is not staffed at all. When selecting a Plans Chief, an optimum choice is one of the initial Incident Commanders because of their understanding of what has already transpired.

Responsibilities of Planning. The planning section is responsible for information gathering and planning at an incident, including the following tasks:

- Supervise the development of the Incident Action Plan (IAP)
- Conduct long range planning and develop alternative strategies
- Develop a demobilization plan
- Collect, evaluate, and display information about the incident
- Determine the need for any specialized resources in support of the incident
- Establish special information collection activities as necessary (weather, environmental, toxins, etc.)

Incident Action Plan. Every incident *must* have an oral or written Incident Action Plan. The purpose of the IAP is to provide all incident supervisory personnel with clear direction and an understanding of incident objectives. Action plans, whether written or oral, are developed for specific time intervals called operational periods. The duration of these operational periods may vary, depending on several factors including the type and complexity of the incident, the projected duration of the incident, the numbers of resources and agencies involved, environmental issues, and safety considerations. Operational periods may be as short as two hours but should not be longer than twenty-four hours.

On smaller incidents, the Incident Commander will usually be solely responsible for developing the incident action plan. However, on larger incidents with more staffing, members of the General Staff and others will contribute to the development of the action plan. The planning section has primary responsibility for documenting the Incident Action Plan, as well as for assembling, printing, and distributing it to appropriate personnel.

While written plans will vary in content and size, some items should be considered for all plans. The major elements of an Incident Action Plan are (Figure 3.23):

- *Incident Objectives.* Clearly states objectives as they apply to the overall incident (ICS 202).
- *Organization.* Describes the ICS organization for a specific operational period (ICS 203).
- *Assignments.* Normally prepared for each Branch, Division, or Group and include the strategy, tactics, and resources to be used (ICS 204).

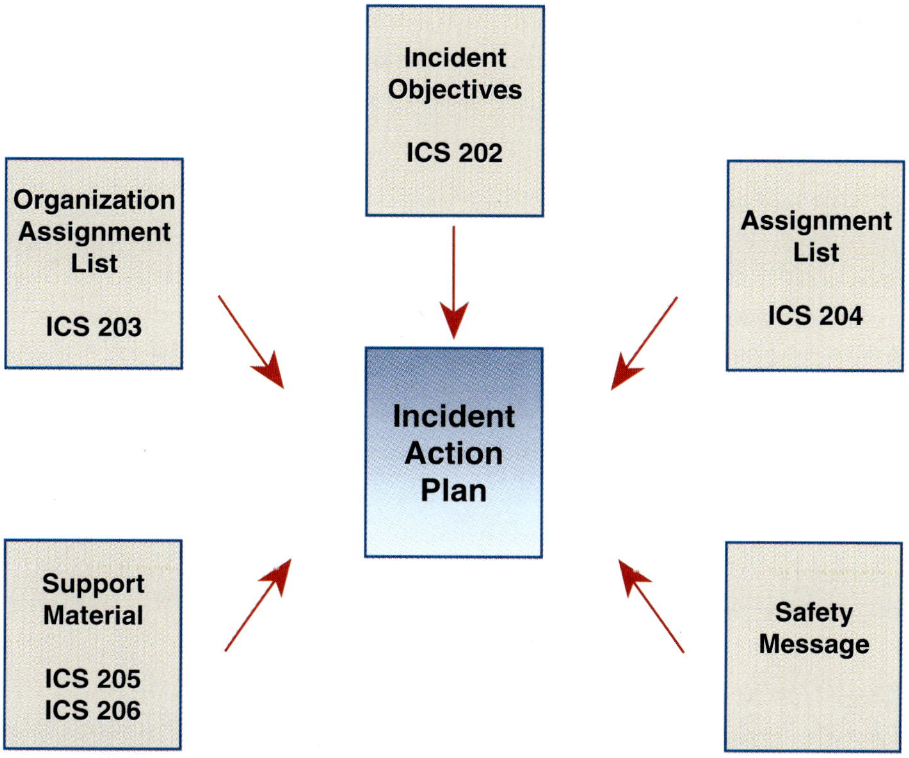

Figure 3.23. Elements of an Incident Action Plan.

- *Support Material.*May include maps, traffic plan, communications plan, medical plan, and similar materials (ICS205, ICS206).
- *Safety Message.* Information/warning regarding issues that might affect firefighter/scene safety.

Planning Units. The Planning Section has five areas of responsibility, or units: Resource Status, Situation Status, Documentation, Demobilization, and Technical Specialists (Figure 3.24). If Command staffs the Planning Section Chief function, all Planning Section responsibilities are realigned and placed under the Planning Chief.

Resource Unit. The Resource Unit Leader is responsible for tracking the status of all assigned resources at an incident. This is achieved by overseeing the check-in of all resources, maintaining a status tracking system indicating current location and status of all resources, and maintaining a master list of all resources committed to the incident. On complex incidents, the Operations Section, as well as Divisions/Groups will also have to track resources assigned to them.

Situation Unit. The Situation Unit Leader may prepare future projections of incident escalation, maps, and intelligence information. The collection, analysis, and organization of all incident information takes place within the Situation Unit.

Documentation Unit. The Documentation Unit (DOC) maintains and safeguards accurate, up-to-date incident files. These may include any

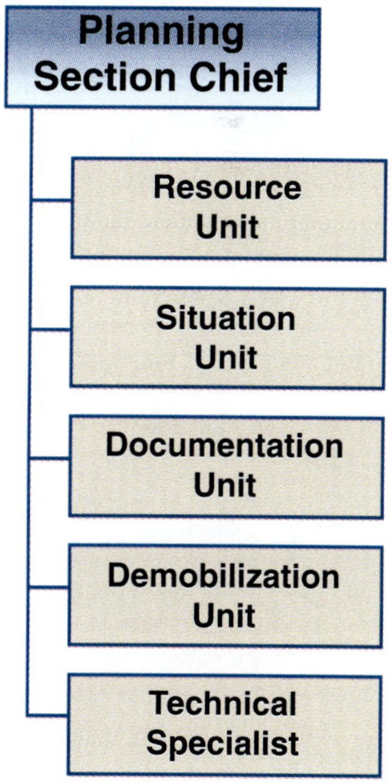

Figure 3.24. Planning-section units.

relevant information from incident reports to injury claims to overtime compensation. The incident files will be stored for legal, analytical, and historical purposes. Duplication services will also be provided by the Documentation Unit.

Demobilization Unit. The Demobilization Unit (DEMOB) Leader is responsible for developing the Incident Demobilization Plan. On large incidents, demobilization can be quite complex and detailed, requiring significant coordination. The orderly, safe, and efficient stand-down from an incident requires appropriate planning. Where costs may be a factor due to the large numbers of varied resources, timely release is important.

Technical Specialists. Not all incidents involve only firefighters. Certain incidents or events may require the use of Technical Specialists (Tech Specs) who possess specialized knowledge and expertise. Technical Specialists may function within the Planning Section or may be assigned wherever their services are required.

Logistics Section

Logistics is responsible for ordering and maintaining essential personnel, facilities, equipment, and supplies necessary to meet the needs of the incident. The Logistics Section Chief is not unlike an army supply officer.

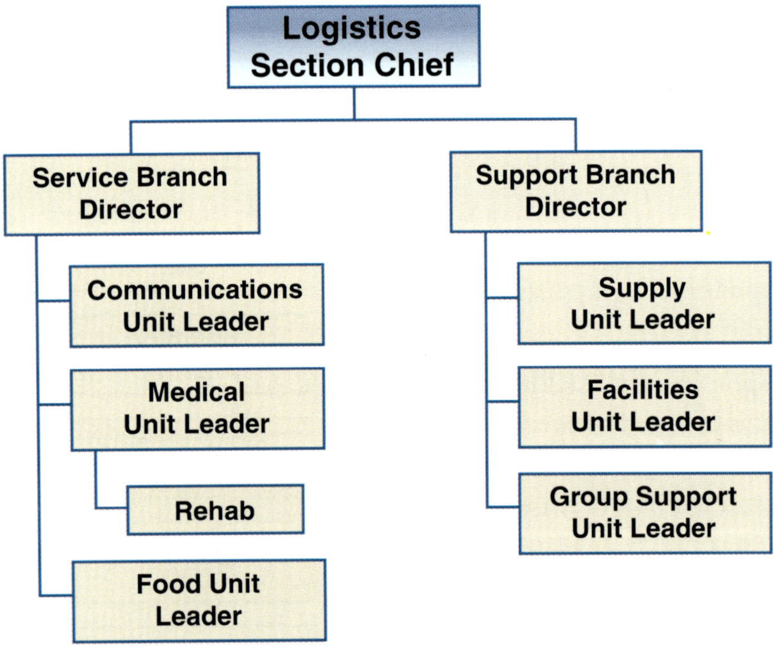

Figure 3.25. Logistics-section organization.

The Incident Commander will determine the need to establish a Logistics Section for the incident. This is usually determined by the size, complexity, and projected duration of the incident. Once the IC has determined the need to establish a separate logistical function, a Logistics Section Chief is appointed. Given an incident of sufficient magnitude, the Logistics Section may be divided into two branches, service and support, with a total of up to six functional units (Figure 3.25). Not all of these units may be required, and they will be established based on incident needs.

Service Branch. The Service Branch is responsible for all service activities. These service needs are met by establishing a Communications Unit, a Medical Unit, and a Food Unit.

Communications Unit. The *Communications Unit* develops plans for the effective use of incident communications equipment and facilities, distribution of equipment to personnel, supervision of the incident communications network, and the maintenance and repair of communications equipment.

Medical Unit. The *Medical Unit* handles the emergency treatment and transportation of injured or ill incident personnel and maintains appropriate documentation. Although the Medical Unit's primary responsibility is to incident personnel, it could be used to assist civilians on the scene should it become necessary. The Medical Unit is also responsible for firefighter rehab.

Food Unit. The *Food Unit* is responsible for feeding incident personnel and providing rehydration fluids. This becomes important in long-term operations or operations in extreme severe weather.

Support Branch. The Support Branch is responsible for developing and implementing logistics plans in support of the Incident Action Plan. The Support Branch Director oversees the Supply Unit, the Facilities Unit and the Ground Support Unit.

Supply Unit. The *Supply Unit* is primarily responsible for ordering personnel, equipment, and supplies. Its additional responsibilities include receiving and storing incident supplies, maintaining supply inventory, and servicing supplies and equipment.

Facilities Unit. The *Facilities Unit* is primarily responsible for laying out incident facilities such as the Base, Command Post, and Camps. This unit also provides sleeping and sanitation facilities for incident personnel and manages Base.

Ground Support Unit. The *Ground Support Unit* is responsible for supporting out-of-service resources; transporting personnel, supplies, and equipment; fueling, maintaining, and repairing vehicles and other ground-support equipment; and implementing a Traffic Plan for the incident.

Finance/Administration Section

The Finance/Administration Section is responsible for all financial, administrative, cost analysis, and legal liability issues of the incident. This position does not have to be filled by a fire officer; however, the person who fills it must be knowledgeable of its responsibilities as well as of agency fiscal policies.

The Finance/Administration Section consists of four units: a time unit, a procurement unit, a compensation/claims unit, and a cost unit (Figure 3.26). Depending upon the incident, any or all of these units may be established as necessary.

Time Unit. The *Time Unit* is responsible for equipment and personnel time-recording. In light of recent court rulings regarding work hours and overtime, and depending on your agency's mutual/auto aid agreements, this could be an important function.

Procurement Unit. The *Procurement Unit* is responsible for handling all financial matters pertaining to vendor contracts, leases, and fiscal agreements.

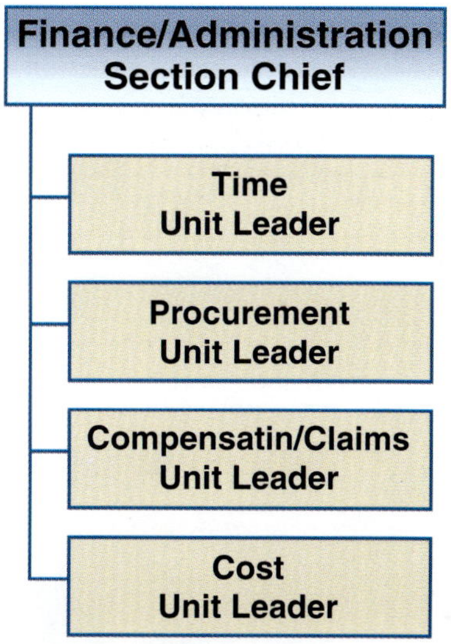

Figure 3.26. Finance/administration-section organization.

Compensation/Claims Unit. The *Compensation/Claims Unit* is responsible for all issues pertaining to compensation for injury and noninjury claims related to the incident.

Cost Unit. The *Cost Unit* is responsible for collecting all cost data, analyzing the data, making cost projections, and recommending cost-saving measures.

Single vs. Unified Command

The incident command system may be structured either as a *Single Command* or a *Unified Command.* In a Single Command structure, only one agency has legal responsibility. Generally this is the way most departments operate on a day-to-day basis. In a Unified Command structure, all agencies with incident responsibility, *either functional or jurisdictional,* jointly provide management direction (Figure 3.27). This *team effort* is accomplished without any agency's abdicating its authority, responsibility, or accountability. This system will allow agencies with different geographic, functional, or legal responsibilities to work together effectively in any given situation. Hazardous materials incidents, mass casualty incidents, natural disasters, wildland urban-interface fires, or any other incident that involves multijurisdictional or multiagency responsibility should use a Unified Command structure.

All agencies with responsibility to manage an incident contribute to the Unified Command process. They jointly determine overall incident objectives, determine strategies, and plan tactical activities. This reduces duplication of effort, ensures maximum use of assigned resources, and enhances overall cost control.

The features of a Unified Command system include:

- A single integrated incident organization
- One operations section chief to direct all tactical activities

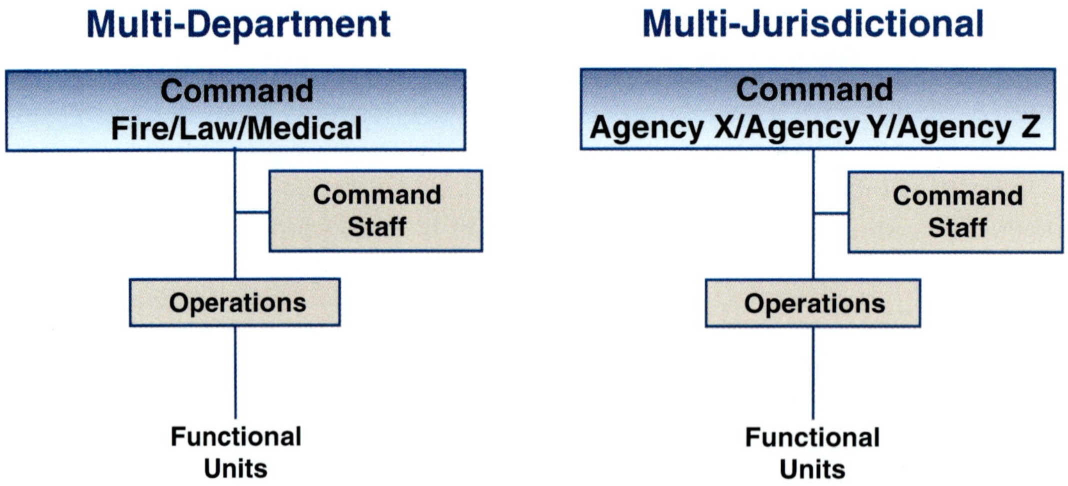

Figure 3.27. Unified-command-system organization.

- Jointly located facilities
- A single integrated planning process and incident action plan
- Shared planning, logistical, and finance/administrative functions whenever feasible

The selection criteria for staffing the Unified Command Post include:

- The location of the incident
- The political jurisdiction involved
- One key official from each jurisdiction or agency
- Representatives from involved departments in a single jurisdiction

Guidelines for the Use of Unified Command

The following guidelines are essential to effectively implementing unified command.

- Understand the Incident Command System's Unified Command. It is critical to success that one understands how the ICS Unified Command functions. A thorough understanding of the principles and structure will ensure acceptance and participation when an emergency requires Unified Command. Joint agency training and education is the best way to avoid "turf battles" and unwillingness to participate. Effectively implementing Unified Command is impossible unless all agencies have committed to the process.
- Collocate essential functions. Failure to do this creates confusion, duplication of effort, and wasted resources. Establish a single Command Post, Planning, Logistics, and other facilities where all involved agencies can operate together.
- Implement early. It is essential to begin joint planning as soon as possible in multijurisdictional or multiagency incidents. This becomes especially important on those incidents where there may be conflicting priorities based on agency responsibilities.
- Agree on an Operations Section Chief and other General Staff positions. Normally the Operations Chief will be from the jurisdiction or agency with the greatest commitment to the incident, although this is not mandatory. However, the Operations Section Chief *should be* the most qualified and experienced person available and the selection *must be* agreed upon by the Unified Command. Since they will have full authority to implement their portions of the Incident Action Plan, the selection of all General Staff members also must be agreed upon by the Unified Command.
- Designate a spokesperson. The Incident Commanders may determine a need to designate one person to be a spokesperson (operation period duty officer) for the Unified Command. This person *does not*

make unilateral decisions for Unified Command but, rather, provides a point of contact as necessary for Command Staff, General Staff, and other appropriate persons.

- Conduct team training. It is important to frequently conduct joint training exercises utilizing Unified Command with adjacent jurisdictions and functional agencies. We tend to "play the way we practice." We cannot wait until an event occurs to exercise our plans and become familiar with policies and procedures.

Generally, the agency with the greatest jurisdictional involvement is assigned the Operations function. Which agencies actually have legal responsibility must be determined depending on the type of incident and its location. Prior experience and training should be considered when staffing the Unified Command Post and the Operations function.

Complex and Area Command

While this text does not specifically deal with incidents of the magnitude that would employ Complex Command and Area Command, it is important for you to have a basic understanding of the various applications and configurations of the Incident Command System.

Complex Command

A *Complex* is two or more separate incidents located in close proximity which are assigned to a single Incident Commander or Unified Command to facilitate resource and incident management. This method of organization has predominately been utilized in a Wildland/Urban urban interface environment; however, it is being employed more frequently in municipal settings when multiple, co-located incidents are encountered.

In addition to wildland incidents, events that might be appropriate for implementing Complex Command could include man-made emergencies such as arson, terrorism, or civil disturbance. Natural disasters, such as floods, earthquakes, tornadoes, or hurricanes, also could present situations where this type of management system should be implemented

Management responsibility for a Complex is assigned to a single incident management team. The organizational level of the individual in charge of each incident within the Complex will vary with resource demand and span of control. Smaller incidents requiring fewer resources may be under a Division Supervisor. As the demand for resources and the span of control requirements increase, one or more of the incidents within the Complex may require a Branch Director. The Complex Command could certainly function adequately even if the individual incidents reached the point where an Operations Section Chief was required; however, at that point the Incident Commander may might want to evaluate the merits of switching from Complex Command to Area Command. When consider-

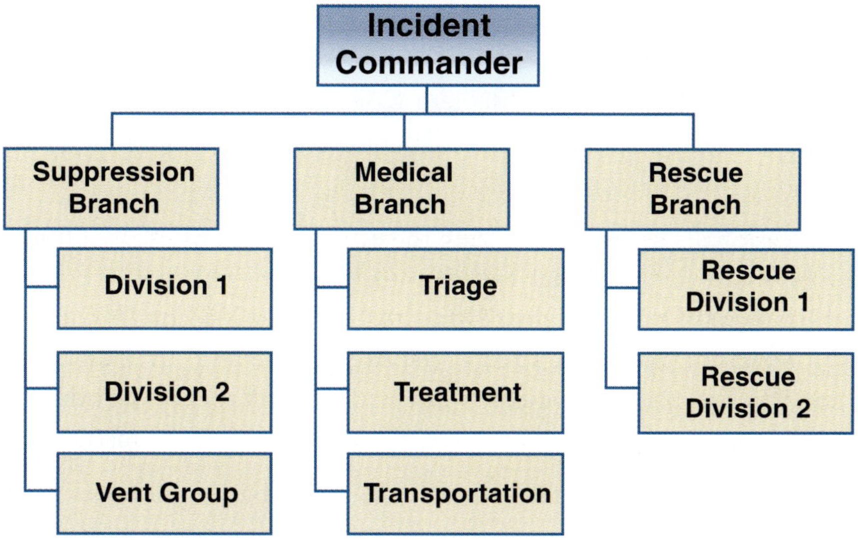

Figure 3.28. Complex command organization.

ing the feasibility of such a move, the IC must not lose sight of the requirements for establishing Area Command and the time involved. Figure 3.28 illustrates a typical Complex Command organization chart.

Area Command

Area Command is an organization established to oversee the management of multiple *like* incidents that are each being handled by an Incident Command System organization, or of a very large incident to which multiple Incident Management Teams are assigned.

Area Command is applicable when incidents in the same general area are vying for the same kind and type of resources (for example, two or more hazardous material spills, structure fires, brush fires, etc.). Due to their size and potential impact, natural disasters, such as floods, earthquakes, and major storms, also should be considered for the possible application of Area Command.

Area Command has been used most commonly for wildland/urban interface fires; however, it also has been utilized in large oil spills, civil disturbances, acts of terrorism, and other large-scale disasters that create multiple incidents and a demand for critical resources.

Establishing Area Command. Since establishing Area Command may take several hours, various Operation Plans should include procedures and criteria for the process. Management personnel should become familiar with these procedures and plans for maximum efficiency, should the need arise.

Area Command is established by the Agency Executive, with the Area Commander's authority often contained in a written Delegation of Authority statement. It is important to note that Area Command *does not* replace the individual Incident Commander's authority, but rather provides an intermediate level of command between Incident Commanders and Agency Administrators. Incident Commanders covered by the Area Command must be notified immediately when Area Command is established.

Occasionally an event requiring the establishment of Area Command may be multi-jurisdictional and/or multi-agency. This will require the establishment of a *Unified Area Command.* When this occurs, full jurisdictional representation is required at the Area Command. In addition, it is essential that all parties have a clear understanding of agency/ jurisdictional expectations, intentions, and environmental and political constraints.

Area Command operates under the same basic guidelines as the Incident Command System (Figure 3.29). This organization, as with all management organizations, should be kept as small as possible while still able to function efficiently. *If necessary,* the Area Commander may establish any positions needed to function effectively, including the following or any other positions deemed necessary to function effectively.

- Area Command Planning Chief
- Area Command Logistics Chief
- Area Command Liaison Officer
- Area Command Information Officer
- Area Command Technical Specialist

Responsibilities of Area Command. For the incidents under its authority, Area Command is responsible to:

- Provide agency or jurisdictional authority for assigned incidents
- Set overall incident-related priorities, ensuring a clear understanding of agency expectations, intentions, and constraints
- Establish critical resource use priorities based on incident need and agency policy
- Allocate critical resources based on these priorities
- Ensure that incidents are properly managed

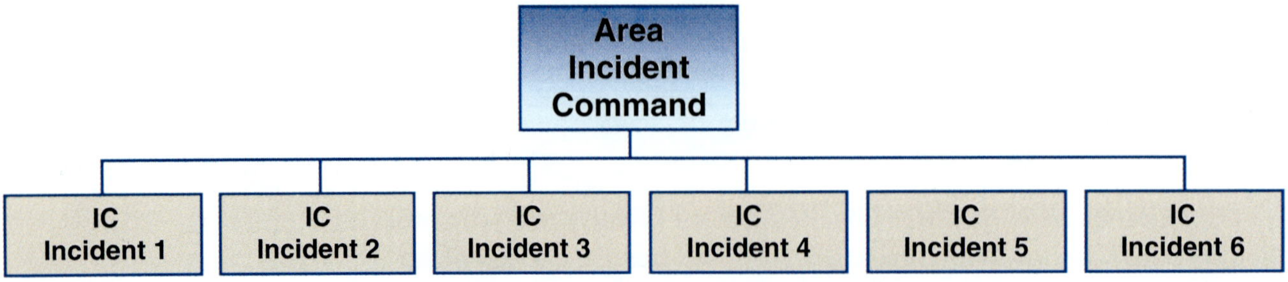

Figure 3.29. Area-incident-command organization.

- Ensure that incident objectives are met and do not conflict with each other or with agency policy
- Maintain communications with officials in charge, with assisting or cooperating agencies, and with other allied groups.
- Coordinate the demobilization and/or reassignment of resources between incidents.

Summary

In this chapter, we have examined the Model Incident Command System in a broad spectrum. The Incident Command System may be used not only on large incidents but on all incidents regardless of size or type. An analysis of various incidents should reveal that nearly all the functions and duties discussed here have been carried out at those incidents. The main point to remember is that all functions must be addressed at every incident. The size and type of incident will dictate the degree to which each function is carried out. How many, if any, of these functional responsibilities can be delegated will be determined by the availability of personnel.

Activity 3.1
Single-Family Dwelling Fire

Consider the following scenario and illustration describing a fire in a two-story, single-family dwelling. Then complete this activity's three parts. First, determine the cues and problems described in the scenario and depicted in the illustration. Second, determine the appropriate strategy for this incident (don't worry about tactics at this point). Third, draw the Incident Command organization chart for this incident. For this activity, use only the following ICS positions:

- Command
- Command Staff (Safety, Liaison, Information)
- Staging
- Division/Groups/Sector
- Single resources

Scenario: You are responding as the chief officer of Battalion 1 to a fire reported to be on the first floor of a two-story single-family dwelling. Engine 1 has arrived on scene confirming a working fire on the first floor of a wood-frame dwelling, with about 20 percent involvement and a confirmed rescue situation. Engine 1's officer has ordered a second alarm. It is 2300 hours.

Resources:

First Alarm: E-1, E-2, E-3, T-1, Sq-1, Medic 1, B/C1

Second Alarm: E-4, E-5, E-6, T-2, B/C2, D/C1, Support Personnel

Activity 3.1

Part 1

Cues and Problems

From the scenario, the graphic and your knowledge of single-family dwellings, list the cues that pertain to this fire situation. We'll start you off with a couple.

Cues: (prediction is an important part of size-up. Besides the obvious, what other cues are present, based on your experience?)

Time of day

Victim visible from window

Problem Identification:

Occupants possibly trapped or overcome

Activity 3.1

Part 2

Strategy

Name the strategies that apply to the first-in company officer and chief at this incident. Use Lloyd Layman's RECEO VS from chapter 2. List strategies by priority of accomplishment.

Activity 3.1

Part 3

ICS Organization Chart

Based on your strategy and projected assignment of resources, complete an ICS organization chart for your operation.

Activity 3.2
Multi-Alarm Incident

Consider the following scenario and illustration describing a downed commuter aircraft that is threatening several structures. Then complete this activity's three parts. First, determine the cues and problems described in the scenario and depicted in the illustration. Second, determine the appropriate strategy for this incident (don't worry about tactics at this point). Third, draw the Incident Command organization chart for this incident.

Scenario:

You are a chief officer responding to a reported crash of a 50-passenger commuter aircraft in a residential area. Three engines, two trucks, and a Medic Unit are on the assignment with you. While responding, you observe a large, dark column of smoke from what you know to be a residential area of the city.

As you approach the scene, you notice the aircraft is pretty much intact but involved in fire. One exposure building has fire showing, and there are several injured persons walking about the area. It is 1600 hours on Thursday.

Resources:

- **First Alarm:** 3 Engines, 2 Trucks, B/C 1, 1 Medic Unit
- **Second Alarm:** 3 Engines, 1 Truck, B/C 2, 1 Medic Unit, Support Personnel
- **Third Alarm:** 4 Engines, 1 Squad, B/C 3, 1 Medic Unit

You are a chief officer responding to a reported crash of a 50-passenger commuter aircraft in a residential area. Three engines, two trucks, and a Medical Unit are on the assignment with you. While responding, you observe a large, dark column of smoke from what you know to be a residential area of the city.

As you approach the scene, you notice the aircraft is pretty much intact but involved in fire. One exposure building has fire showing, and there are several injured persons walking about the area. It is 1600 hours on Thursday.

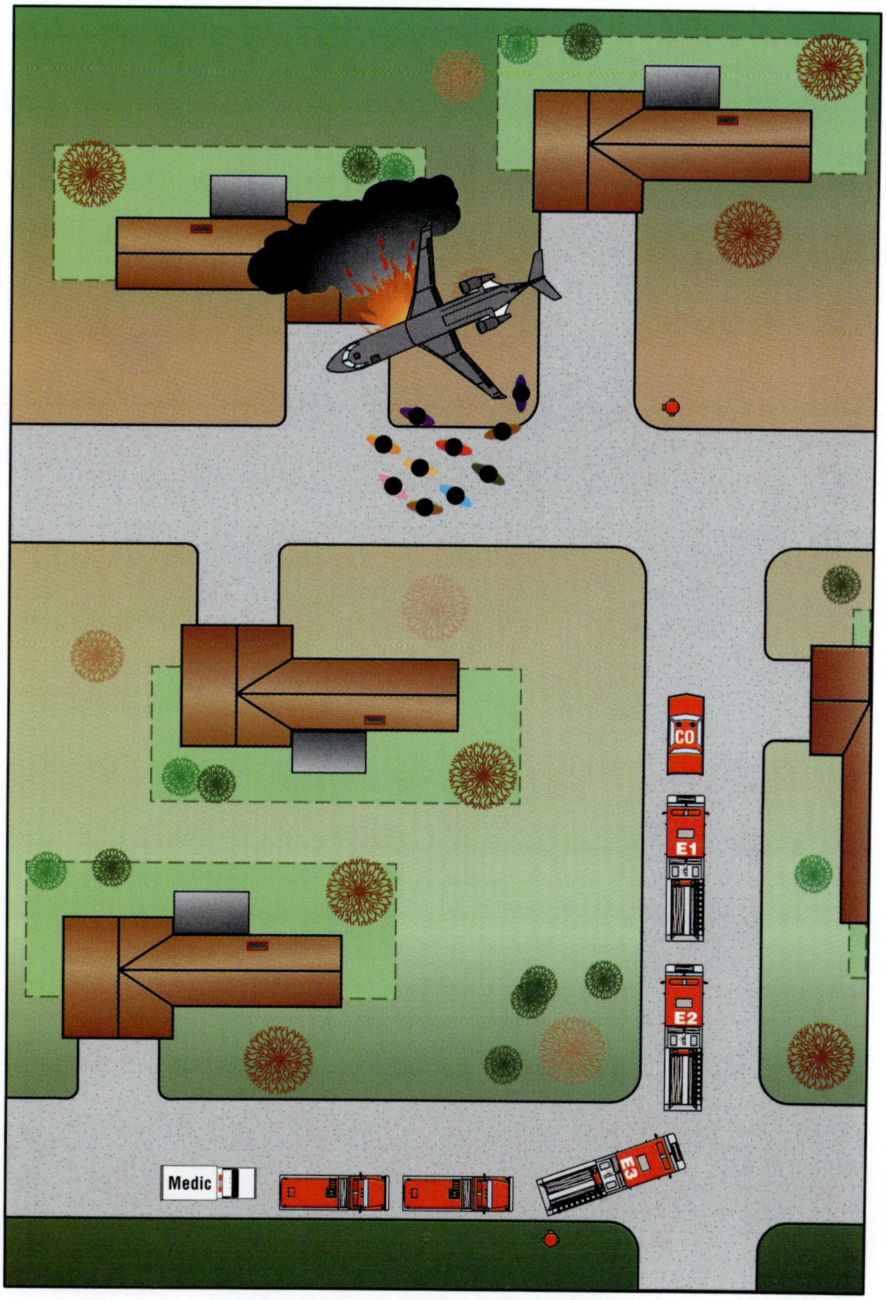

Activity 3.2

Part 1

Cues and Problems

From the scenario, the graphic, and your knowledge of multialarm incidents, list the cues that pertain to this situation. We'll start you off with a couple:

Cues:

Aircraft: Mass Casualty

Building Fire: Life Hazard

Problem Identification:

Multiple Fires

Activity 3.2

Part 2

Strategy

Name the strategies that are applicable to the *first-in company officer* and *first-in chief* at this incident. Use Lloyd Layman's RECEO VS from chapter 2. List strategies by priority of accomplishment.

Activity 3.2

Part 3

ICS Organization Chart

Based on your strategy and projected assignment of resources, complete an ICS organization chart for your operation.

Suggested Activity Answers

Activity 3.1
Part 1: Cues and Problem Identification

Cues:
Time of day—high occupancy load

Victim visible at second-story window—confirmed rescue

Other victims probably still in dwelling

Smoke from first floor window—probably in family room

Fire could have started in basement

Fire probably spreading horizontally

Car in driveway—reaffirming occupancy

Age and design of house—probably platform construction

Problem Identification:
Possible that fire has blocked stair egress from second floor

Possible occupants overcome on second floor

Possible extension into attic space

Probable that rescue will have to be effected from exterior

Strategy:
Rescue and primary search of the occupancy

Confinement—secure the stairs, check second floor and adjacent rooms, and basement

Ventilation—relieve smoke from both floors and attic if necessary

Extinguishment—putting out the fire

ICS Organization Chart:

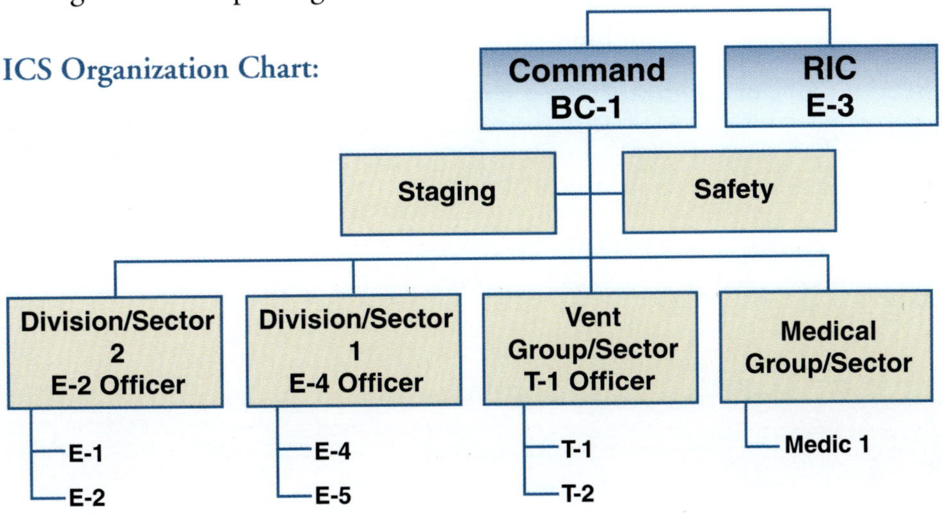

Activity 3.2
Part 1: Cues and Problem Identification

Cues:

Aircraft—presents mass casualty

Jet fuel—requires special extinguishing agents

Smoke and flame from exposures—multiple fires

Passenger capacity and secondary fires—resource intensive

Aircraft intact—viable rescue

Problem Identification:

Fire in both aircraft and exposure building

Aircraft interior fully charged with smoke

Occupants of aircraft possibly trapped/overcome

Second floor of exposure building charged with smoke

Occupants of building possibly injured/overcome

Possible extension into attic of exposure building

Environmental impact of fuel (sewers, creeks, etc.)

Possible jet fuel fire

Possible explosion hazard

Possible fire extension to other exposures

Evacuation of exposure buildings

Large numbers of victims requiring treatment/transportation

Need quantity of foam

Strategy:

Rescue—aircraft and building, including primary search of building

Exposure protection—protect passenger cabin of aircraft; check for extension of fire in building; check for extension of fire in other exposure buildings

Confinement—protect rescue efforts in aircraft; check attic in fire building

Extinguishment

Ventilation—both aircraft and building

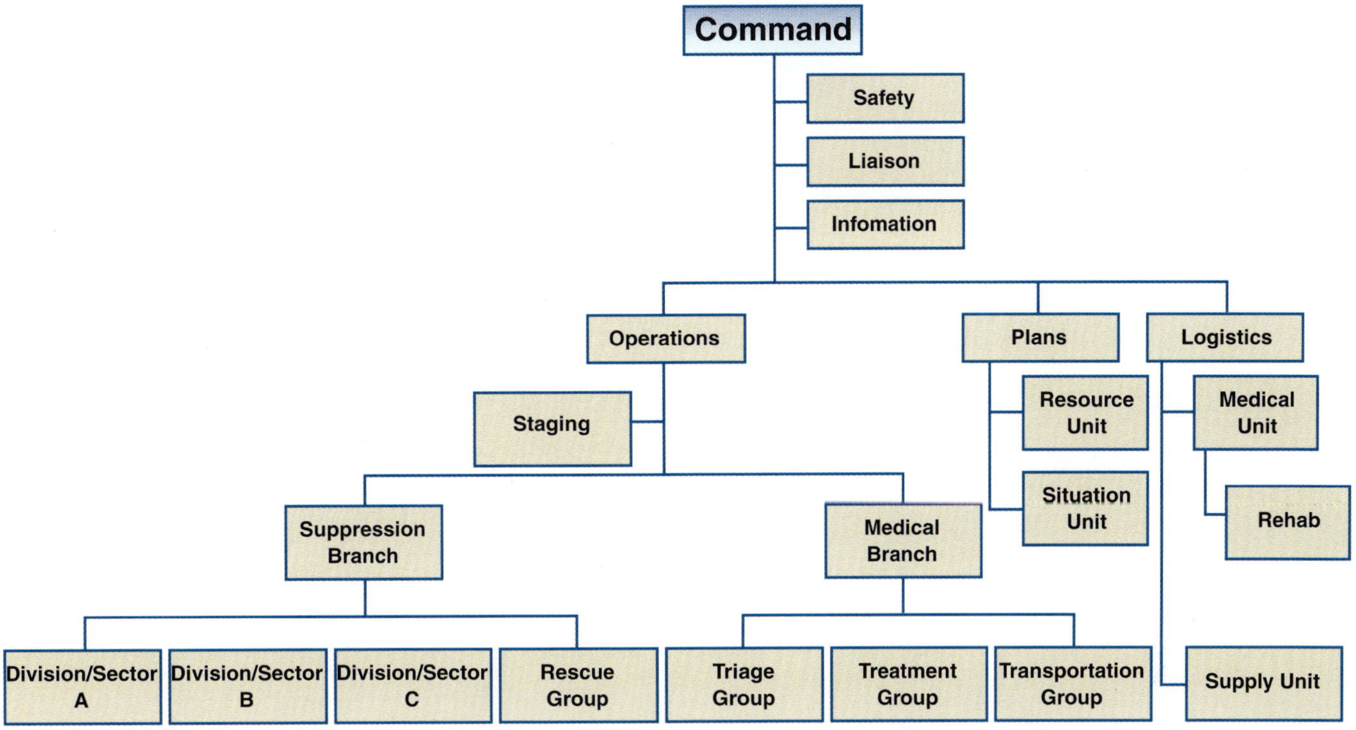
Command
Safety
Liaison
Infomation
Operations
Plans
Logistics
Staging
Resource Unit
Situation Unit
Medical Unit
Rehab
Suppression Branch
Medical Branch
Division/Sector A
Division/Sector B
Division/Sector C
Rescue Group
Triage Group
Treatment Group
Transportation Group
Supply Unit

Chapter 4
Garden Apartments

Garden Apartments

Chapter 4

Introduction

Garden apartments are found throughout most of the country. These housing units are cost effective for builders and developers and allow them to attain a reasonable profit. They have seen a boom on the East Coast and in mid-America, with numerous units being built in the West also.

Garden apartments are three- or four-story residential structures (Figure 4.1). Usually, they have only one exterior entrance with a common set of stairs to the upper floors, although some may have two entrances and stair shafts. The norm is four apartments per floor. Often, the first floor is set half-way into the ground, requiring a walk down a half-set of steps. Also, ground contours often allow the rear of the first floor to be at ground level so that a sliding glass door and patio can be installed.

Figure 4.1 Age and physical ability of occupants may dictate strategy.

A number of construction types are used in garden apartments, including wood-frame with vinyl siding, aluminum siding, or brick veneer; ordinary (masonry, wood-joist) construction; and, rarely, noncombustible construction with steel bar-joists supporting concrete floors and concrete block walls.

Wood-Frame Construction

Wood-frame construction comes in three modes: (1) platform, (2) balloon-frame, and (3) post and beam.

Platform construction is the modern methodology that starts with a foundation. A floor assembly is built on the foundation. Then, the walls are erected up to the level of the next floor or the roof assembly. If the next level is a floor, then the floor is erected and a second set of walls are constructed, and so forth, up to the level of the roof assembly. The walls are load-bearing members. This type of construction provides natural fire stopping at each floor level, and all the lumber can be held to an eight- to ten-foot length for the walls. The walls in this type of modern construction are four inches thick (some may be six inches if the builder is concerned about energy loss).

Balloon-frame construction is a dated methodology that was used from around the beginning of the twentieth century until about the 1940s. In balloon-frame construction, a foundation was constructed, and the walls were erected using long lumber (usually two-by-fours). The walls were uninterrupted all the way to the attic level, even in multistory construction, usually to a maximum of two-stories. Ledger boards were then nailed to the walls at each floor level; the floors' beams were laid on the ledger boards; and then the floors were finished. The walls are load-bearing members. This type of construction provides no fire stopping from the basement to the attic, and all void spaces in the wall and floor assembly are interconnected. Thus, fire can spread very rapidly in these buildings if it enters a wall or floor void space.

Post and beam construction is a building methodology found in many barns and sheds. Here, large columns (posts) and beams (four-by-four or greater) are erected and nailed or bolted together. Wooden floors are erected within the posts and beams. Finally, wooden planks are placed on the outside to create the walls. Obviously, the walls are not load-bearing members.

Ordinary Construction

In ordinary construction (also called masonry, wood-joist construction), footers are dug and a foundation laid for a concrete-block wall assembly. The walls are erected up to the roof-line level. The floor beams are put into place on ledger boards or in sockets in the walls. A column system is normally found in the basement or crawl space. The columns carry the floor beams over long stretches to opposite walls. Interior bearing walls located over the columns carry the loads from the upper floors. Usually the floors are finished with sheathing and hardwood flooring. The walls in this type of construction typically range from 8 to 12 inches thick.

Noncombustible Construction

In noncombustible construction, none of the bearing members may be combustible. This means they are usually made of concrete and/or metal. Footers are poured and walls are erected. The walls may be concrete block, poured-in-place concrete, tilt-up (concrete slabs set in place), or metal prefabricated units with or without windows. These buildings often require an engineered steel skeletal assembly, depending on the construction technique used. The next floor up either is poured-in-place concrete, with or without steel bar-joists, or consists of prestressed or poststressed concrete assemblies. Then another set of walls is installed. Any interior partitions must have noncombustible studs (metal).

The cues found in garden apartments are also applicable, in varying degrees, to:

- Center corridor apartment buildings
- Rooming houses
- Motels
- Townhouses
- Condominiums

Occupancy-Specific Cues

Wall Construction

The bearing walls in garden apartments will be one of two basic construction types. They may be masonry walls of concrete block with vinyl or aluminum siding or brick veneer, or they may be wood-frame with vinyl or aluminum siding or brick veneer.

Garden apartments usually have four apartments per floor. Thus, a three-story garden-apartment building will have 12 apartments, a four-story building 16. We'll call this package of 12–16 apartments a *section*. Often, two or more sections adjoin one another. New codes usually require fire walls or fire division walls between sections.

The partition walls are usually half-inch drywall (gypsum board) on wood or metal studs placed 16–24 inches apart. Remember, *you can go through a flimsy wall very quickly* to get to the opposite side of a fire area.

Roof Assemblies

Under fire conditions, roof assemblies can collapse and injure or kill firefighters operating on their decks or on the floors below. *You must know the type of roof assembly onto or under which you are sending your personnel.*

Flat Roofs. Many garden apartments have flat roofs (Figure 4.2).Flat roofs are normally constructed in one of the following methods:

- *Beam and rafter with sheathing.* This assembly is normal for older flat-roof construction. It uses horizontal beams, usually ranging from 2 inches by 6 inches up to 2 inches by 12 inches. To these beams are attached roof rafters on 16-inch centers. A sheathing (decking) is laid over the rafters. Depending on when the building was constructed, the sheathing may be 1 inch thick and made of diagonally laid boards or sheets of plywood or particle board. This very strong assembly tends to burn through rather than collapse. It is an ideal roof for ventilation activities during fire fighting.
- *Steel bar-joist/poured concrete/prestressed or poststressed concrete.* Found in newer complexes where noncombustible construction is used, this assembly is obviously not one that can be easily ventilated or penetrated to assist in an attack on an attic fire.
- *Parallel-chord wood truss or plywood I-beams.* Found in modern construction, these lightweight building materials are widely used today. They are strong and cost effective for builders, but unfortunately under fire conditions, they have been known to fail en masse in as little as five minutes after exposure to flames. Ventilation tactics on such roofs must be approached with great caution. A number of firefighters have been killed when this type of roof has collapsed from under them or collapsed onto them from above.

Figure 4.2 Tar-and-gravel or asphalt roofing are the most common exterior coverings for flat roofs.

Pitched Roofs. Some garden apartments have pitched roofs (Figure 4.3). Pitched roofs typically are of either ridgepole-and-rafter construction or wood-truss construction.

- *Ridgepole-and-rafter construction.* Although ridgepole-and-rafter construction is still used where a usable attic space is desired, it typically indicates older construction. Rafters are angled upward from the wall line toward the center of the space. They are attached to a piece of lumber running horizontally, which is the ridgepole. The meeting of the rafters from two walls at the ridgepole creates a pitched roof assembly. This assembly, like the flat beam and rafter assembly, is very strong and tends to burn through rather than collapse over a wide area.
- *Wood-truss construction.* Wood truss is the modern roof assembly base. A truss assembly is constructed of wood or metal members. Two chords span the horizontal length of the truss, one on the top and one on the bottom. Short members are then placed at an angle between the chords, forming the *web* (Figure 4.4). In plywood I-beams, the top and bottom chord are perfectly parallel, and the web is a solid piece of plywood or strandboard. In a pitched truss, the top chord is angled to form a peak. Wood trusses can be nailed at the

Figure 4.3 Exterior coverings on pitched roofs vay and may affect ventilation efforts.

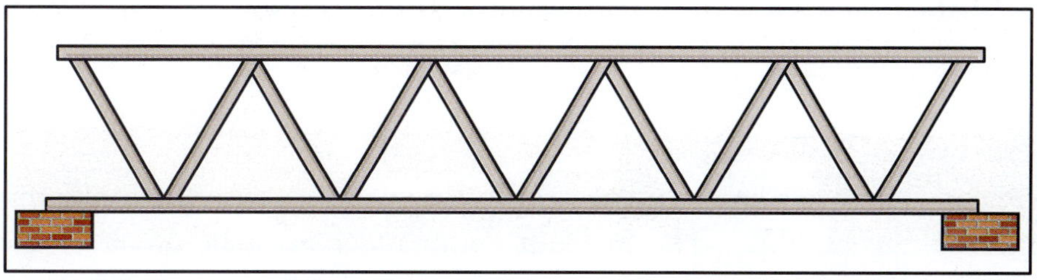

Figure 4.4 Wood truss construction.

joints. Toe-nailing the joints was the original method of truss assembly, but to reduce labor costs, speed construction, and strengthen the truss assemblies, manufacturers started using gusset plates. A gusset plate is simply a piece of plywood or sheet metal that is laid over the truss at the joints. Nails are then driven through the gusset plate and into the truss members.

The latest truss assembly technology is the gang-nailer. A gang nailer is a piece of thick sheet metal that has been punched to produce a large number of ⅜-inch points protruding from one side. The gang-nailer is laid on the truss at the joints and pressed into the truss members to hold them together. Any truss assembly that has been exposed to flames or high heat build-up is dangerous for firefighters. Ventilating directly over the heated or flame-impinged area can have devastating effects. Firefighters may be able to ventilate these assemblies in areas that are not yet significantly affected by the heat build-up—for example, toward the rear of a building when the fire is toward the front of the building. The incident commander must have the knowledge and experience to accurately predict fire conditions in the building.

Attics and Cocklofts. When a garden apartment has no fire walls or division walls, it normally will have a common attic or cockloft over the entire structure. These typically include two or more sections. Even when there are fire walls, there is a common attic over the top four apartments in a section.

The common attic or cockloft is an avenue for rapid horizontal fire spread down the building. It is critical that the Incident Commander know the type of construction and realize that it has a common attic. Crews must be sent to the upper floors when fire is suspected to be in the attic space. These crews must pull ceilings and inject water to cut off the fire. *Vertical ventilation is critical to success in attic or cockloft fires.*

Access

It may be difficult or impossible to drive fire apparatus to the sides or rear of a garden apartment, or even to get close from the front. Resident parking may block parking lot entrances or preclude the movement of fire apparatus through the parking lot. The experienced fire officer knows that apartment complexes often have more cars than apartments. Residents and guests park in many places where they should not. Proper planning of an apartment complex needs to take this into account.

Entry

Making entry to a garden apartment section will vary with the configuration of the building. The incident commander must be aware of the construction methods used for these complexes in the response area.

Figure 4.5 Often a common entrance will serve several units.

Front Entry. Usually a single front access door serves the common open stairwell of each section in a garden apartment (Figure 4.5). Each apartment has a single entrance door from the common stair and vestibule area. The apartment door is typically steel or steel-clad; however, many are wooden. The stairs are often made of metal and are self-standing. However, they also may be made of wood. The apartments may have balconies served by sliding glass doors in the front on second and higher floors. These balconies may serve as entrance points for firefighters since the stairs may become crowded with fleeing occupants and entering firefighters. Rescue from the balconies can be a major problem at serious fires in garden apartments.

Rear Entry. Windows usually are plentiful at the backs of garden apartments. Sliding glass doors may or may not be present at ground level, and the first floor of most garden apartments is partly underground. Many times the land is contoured to accommodate a patio area served by sliding glass doors in the rear first-floor units. Balconies served by sliding glass doors may be present at the rear on the second and higher floors.

Floor Assemblies

Older garden apartments have beam-and-rafter floor assemblies with plywood sheathing. Apartments built of noncombustible construction typically have steel bar-joists with poured concrete floors. Floor assemblies in modern construction may be beam and rafter, parallel chord wood-truss, or parallel chord wood I-beam with plywood sheathing.

Storage Rooms

A ground-level apartment space may be used as a storage area for one or more sections of the apartment complex. These storage areas provide a small cubicle for each apartment owner. Normally, the cubicles are constructed of plywood and may have chicken-wire roofs. It is possible to find anything in these cubicles, from flammable liquids to motorcycles to pesticides and gunpowder. This area may also contain several clothes washers and dryers. Due to the varied contents of the storage area and the possibility of mischievous acts by residents or visitors, the storage area is a likely place for a fire.

Common Problems

Always consider that every apartment is occupied 24-hours a day and remember that unattended children may be left in any apartment. Some garden apartment complexes cater to the elderly, and these need to be identified.

Rescue, including primary search, is going to be the most labor intensive action required at a garden apartment complex. From forcing doors to doing a crawling search, time will work against the operating crews and the IC. The Incident Commander should recognize early in the situation what requirements the operating forces will have to meet. The IC also should make an early call for more resources in the form of a second or third alarm or mutual aid.

Along with the major search and rescue effort, a major medical assignment is needed. A garden-apartment fire has the potential to be a multicasualty incident as well as a fire incident. The IC should call for sufficient medical resources on the first indication of injuries. Even a simple fire in a pot of food on the stove with an evacuation may cause an elderly person to have a heart attack or chest pains or some other ailment. At a minimum, one EMS support unit should be dispatched with the first alarm assignment.

Roof or Floor Assembly Construction Type

Prefire planning *must* determine ahead of time whether a specific complex has wood-truss roof and/or floor assemblies. A command officer has no time on arrival at a fire incident to determine the exact type of roof or floor construction.

Truss roof assemblies tend to be more stable in garden apartment structures than in structures without partition walls. Partition walls are not bearing members, but they do support the bottom truss chord and reduce collapse potential. Nonetheless, ventilating a truss roof can be hazardous to your firefighters if you have not done an effective risk-benefit analysis. When the truss appears to be involved, either be sure to have firefighters ventilate far enough away from the burning area to be safe or do not ventilate at all. As the IC, you must be the judge.

When a beam and rafter or ridgepole and rafter roof assembly is involved, firefighters should avoid the area directly over the fire, but areas adjacent to the burning area are more than likely safe for operations. These roof types tend to burn through rather than collapse en masse.

Finally, remember that if you won't put personnel on a roof, then you also shouldn't allow them to operate under it.

Fire Involvement Considerations

The Command Officer must read the fire situation and predict what else is happening that he or she cannot see. Prediction is an important part of size-up. The IC must be able to project where a fire is inside a structure, where it is going, how intense it is, and what it (or toxic products of combustion) will probably threaten if it moves unchecked in any direction. The IC must read the apparent cues and use them to predict the other probabilities.

How will you make predictions? From experience—either from real incident scenes or gleaned from others in training, journal articles, or books. You learn as much as possible from every source available. You do not get experience sitting around the kitchen table at the station complaining about the department officers and chief. You get experience by working at it, a little every day, week after week. As one good fire service officer said, "You train your brain."

Let's look at some more probabilities and their cues in garden apartments. Has the fire extended to the roof assembly area? What are the signs of probable roof assembly involvement?

- *Major fire involvement in one top-floor apartment.* With a well-involved apartment, the fire may have entered ductwork through the grills and extended into the walls and possibly into the attic space.
- *Fire involvement in two or more top-floor apartments.* With two apartments on fire, the ceiling material likely has failed and the fire probably is directly active in the attic space.
- *Hot, rising smoke from attic vents.* This may only be smoke in the attic, but treat it as if there is a fire up there.
- *Fire has vented through the roof.* The fire definitely has reached the attic space.

Is the fire likely to spread to the roof assembly in the near future? What are the signs that the fire will likely spread to the roof or ceiling assembly?

- *A well-involved single occupancy on the top floor.* Even without further signs of attic involvement, the fire is likely to extend into the attic space in the near future. Tactics for attic involvement should be employed.

Structural Deterioration and Collapse

Garden apartments are not prone to early collapse. However, should fire gain a good foothold in a combustible floor or roof assembly, the likelihood of failure or collapse increases proportionally. Older buildings may have structural deficiencies before the fire.

Cue-Based Predictions

By applying the occupancy-specific cues in a garden-apartment fire situation, we can make a number of predictions and assign groups/sectors accordingly.

Rescue Group/Sector

The life hazard could be severe, and a thorough primary search must be done. This occupancy requires a specific assignment for search and rescue. You should establish a Rescue Group or Sector to handle the primary search and rescue of occupants. You may need to assign from two to four rescue personnel per apartment, depending on smoke conditions on each floor and in each apartment. This certainly justifies having one Rescue-Group or Rescue-Sector Supervisor responsible for all search and rescue in the building. Calling for additional resources as early as possible may mean the difference between no loss of life and a disaster. Don't forget that an EMS assignment will be needed.

Rescue Group/Sector

Medical Group/Sector

Smoke conditions in the stairwell and corridor may be heavy and prevent occupants from exiting. When lower floors are involved, smoke is likely to enter the stair shaft and travel up. This will prevent occupants from escaping on their own. Often, if balconies are available, occupants will go to them and await ladder rescue. It may not be necessary to get everyone off of every balcony, depending on the fire's location and the smoke conditions on the balcony's side of the building. At garden apartments without balconies, you will find occupants at various windows in their apartments. Due to smoke build-up in certain apartments, you may have a real rescue problem on your hands.

Medical Group/Sector

Exposure Group/Sector

Fire travel in attic spaces could involve an entire section. Once fire enters the attic space, that entire section, or maybe even additional sections, will be threatened. Roof ventilation is critical, along with the placement of hose lines on the top floor. Ceilings should be pulled and streams directed to extinguish the attic fire as soon as possible. If other sections are exposed, the IC should establish an Exposure Group or Exposure Divisions, or Exposure Sectors.

Exposure Group/Sector

Division/Sector 2

Fire in a storage area may spread rapidly to the floor above and may involve unsuspected materials. These materials may include anything under the sun, from the simplest item to the most exotic. Often the ceilings of storage rooms are unfinished, which leaves the floor assembly for the second floor exposed from below. A working fire in the storage room can quickly penetrate the floor above and cause the apartment unit to become involved in fire. At the very least, heavy smoke conditions will prevail in the apartment above the fire. It is critical to get crews to the floor above and check for extension. In ICS terms, this would require establishing a Division 2 or Sector 2

Division/Sector 2

Incident Management Cues

Coordinated Operations

Rescue and primary search must be coordinated with ventilation. Life safety is the highest priority, and early ventilation efforts should support rescue. Rescue and primary search teams must take hand lines with them, or other crews must support them by staffing appropriately placed lines.

Where and how do you search a space such as an apartment? First, every department should follow a clearly mandated search method. The two basic methods are left-hand searches and right-hand searches. In a left-hand search, search personnel enter the space at the door and turn left. They make every left-hand turn possible as they follow the wall around the perimeter of the room. Of course, where only a right-hand turn is possible, they take the right-hand turn. This will eventually lead them back to their entry door, regardless of the number of rooms they search in the process. It is important for personnel not to lose contact with the wall. If they do lose contact, they can get disoriented and lost and not know how to get out. Therefore, they should take a tool such as a short pike-pole to sweep the floor away from the wall and under furniture and beds. The right-hand turn search mirrors the left-hand turn search: personnel turn right at the entry door and make every right-hand turn possible. Each department must adopt only one of these methods and train on it appropriately.

It is generally best to begin a search on the fire floor, working from the fire outward. Next, personnel should search the floor directly above the fire floor, then the structure's top floor. Finally they should work their way back down from the top floor, one floor at a time. This rule of thumb is not fixed in granite; the underlying principle is that rescuers should search the locations where people are in the greatest danger and work their way from there.

Searching is resource intensive. You must place sufficient personnel in the Rescue Group/Sector to complete the job before people die, and you should assign them to search in pairs. This does not mean they must be side-by-side, but they should be in the same proximity. They should always know each other's location and where they are going, and they must have a specified place and time to meet. If one gets to the meeting point and the other is not there, the first will wait a predetermined time and then go looking for the second after notifying the supervisor.

Confinement efforts are important, and coordinated ventilation to assist them is a necessity. Ventilation personnel should open windows and doors that allow attack and confinement crews to push the fire from the unburned area to the burned area and out of the building, if possible.

ICS Organization Cues

Each function (e.g., ventilation and rescue) requires a Group/Sector Supervisor. Each geographic area where personnel are operating (e.g., Division/Sector 1, Division/Sector 2) also requires a Supervisor. At working incidents, you will need a Safety Officer and an Information Officer. Figure 4.6 provides an ICS organization chart for a typical three-story garden apartment fire with the fire on the first floor.

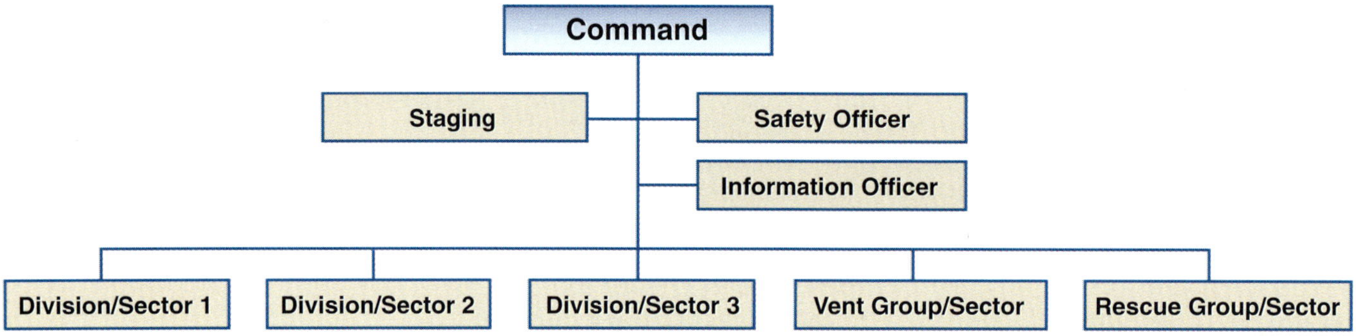

Figure 4.6 ICS organization at a garden apartment fire.

Strategy and Tactics

Any Fire Situation—Primary Search and Rescue

- The IC should establish a Rescue Group/Sector to handle this operation.
- Be sure to check the rear of the building immediately after arrival. The life safety problem may be greater there than in the front. This is especially possible when there are rear balconies. It is not uncommon to find a large number of people on balconies. An early check of the balconies at the rear of the structure will give you a sense of the magnitude of the rescue problem. For this operation, you may need a Division/Sector C.
- First, immediately begin a primary search of the occupancy or occupancies involved in fire if it is possible to enter the area. It is critical that the people who may be in greatest danger are attended

to first. Use hose lines, if necessary, to get as close to the fire area as possible.

- Second, primary search the rest of the apartments on the fire floor. Start from the fire apartment and work out and away from the fire area. The floor where the fire is occurring will normally be charged heavily with smoke. The occupants of adjacent apartments on that floor may need help escaping the toxic materials.

- Third, primary search the floors above, starting with the apartment over the fire and working outward. The floor above the fire is generally charged with from moderate to light smoke conditions. The apartment directly above the fire area will have the greatest need for an early search.

Fire Not Involving the Roof Assembly

- Make an aggressive interior attack on the fire, accompanied by horizontal ventilation. Positive-pressure ventilation may be appropriate here. Getting lines on the fire as soon as possible is critical. Quick action with hose lines will reduce the probability of extension and severely limit the production of toxic products of combustion. Use positive pressure ventilation only when you are sure that no victims are between the fire and the smoke exit point from the building (probably a window). For example, if the fire is on the second floor, then you would very quickly establish a Division/Sector 2 and a Division/Sector 3.

- Assign personnel to check for extension.
  - — Other apartments on the same floor: Division/Sector 2
  - — Apartments on floors and/or attic space above and floors below, in that order: Division/Sector 3

Personnel checking for extension should have hose lines and forcible entry tools and should immediately extinguish any fire they find in walls or ceilings.

Fire in a Storage Area

- Rapidly ventilate the storage area by removing window panes. Positive pressure ventilation may be effective. Relieving the pressure inside the storage room will enhance extension control operations. The open windows also allow a means to attack the fire, depending on its location. The entrance door to the storage room is probably the best attack position since an attack from the doorway protects the stair shaft.

- Aggressively attack the fire with at least two hand lines, one of which is a backup line. This would require a Division/Sector 1.

- Check for extension in the apartments next to and above the involved area. This would require a Division/Sector 2.

Fire in an Attic Space over a Single Section Separated from Other Occupancies by Fire Walls or Division Walls

- If you do not know the type of roof construction, *assume that it is lightweight.* This is critical to firefighter safety and survival. Your department should have a prefire plan for these types of buildings. Lacking such a plan, you must assume the worst case situation, and that is lightweight truss construction for the roof and floor assemblies. Based on our knowledge of failure time for a lightweight truss with flame impingement, caution in the placement of personnel to open the roof or to work under it is essential.
- When the risk is acceptable, the following tactics are recommended:
  - — Interior fire attack: assign several hand lines to the top floor, pull ceiling, and attack the fire.
  - — Perform vertical and horizontal ventilation.
  - — Check interior exposures for extension.
  - — Check exposed attic areas in adjoining sections for extension.

  These actions call for establishing a Division/Sector 3, Vent Group/Sector, and an Exposure Group/Sector.
- When the risk of attacking the fire from the floor below is unacceptable, the following tactics are recommended:
  - — Vertical ventilation, only if it can be done safely away from the intensity of the fire.
  - — DO NOT put a ladder pipe or other exterior stream into any ventilation hole.
  - — Exterior fire attack: heavy streams from front and rear. Use aerial devices to apply streams to upper floors and attic space.
  - — Check interior exposures and attic areas on the far side of fire walls for extension.

Well-Involved Fire in a Roof Assembly that Serves More than One Occupancy (Common Attic or Common Cockloft)

- Roof construction *must* be known *before* an interior attack or vertical ventilation is initiated. If roof construction type is not known, *assume lightweight.*
- Risk/benefit must be assessed before personnel are placed on or under a roof or floor assembly of truss construction that is well-involved in fire.

— Vertically ventilate as close to the fire as safety permits. When possible, the vent hole should be on the side with the shortest roof span to the end of the building or a fire wall. Start by trying to draw the fire toward the least amount of building.

— DO NOT put a ladder pipe into any ventilation hole.

— Fire attack with hand lines from beneath must be aggressive. Confinement of the fire to prevent spread to the longest portion of the structure is a high priority.

— Check interior exposures for extension.

— Pull ceilings to check exposed roof areas for extension and provide a path for water to reach the fire.

These incidents require the same ICS organization as a fire in an attic space over a single section separated from other occupancies: establish a Division/Sector 3, Vent Group/Sector, and an Exposure Group/Sector.

Activity 4.1

Garden Apartment Fire

Directions

Read the scenario of a fire situation and examine the accompanying plot/floor plan. Then assume the role of the first-in chief officer at the incident and complete this activity's four parts. First, determine the cues and the problems that are presented in the scenario and the plot/floor plan. Second, determine the strategy and tactics for the incident. Third, specifically identify the tactics or tasks that will be performed by the resources you call to the scene. And fourth, draw an Incident Command System organization chart for the operation.

Scenario

You are responding to a garden apartment fire as a Chief Officer. There are three Engines, a Ladder Company, and a Medic Unit on the assignment with you. As you approach the scene, you see heavy smoke and flame exiting a broken window on the third floor in an area that should be a bedroom. It is 2300 hours. The local fire alarm has been pulled, and some occupants are exiting the building.

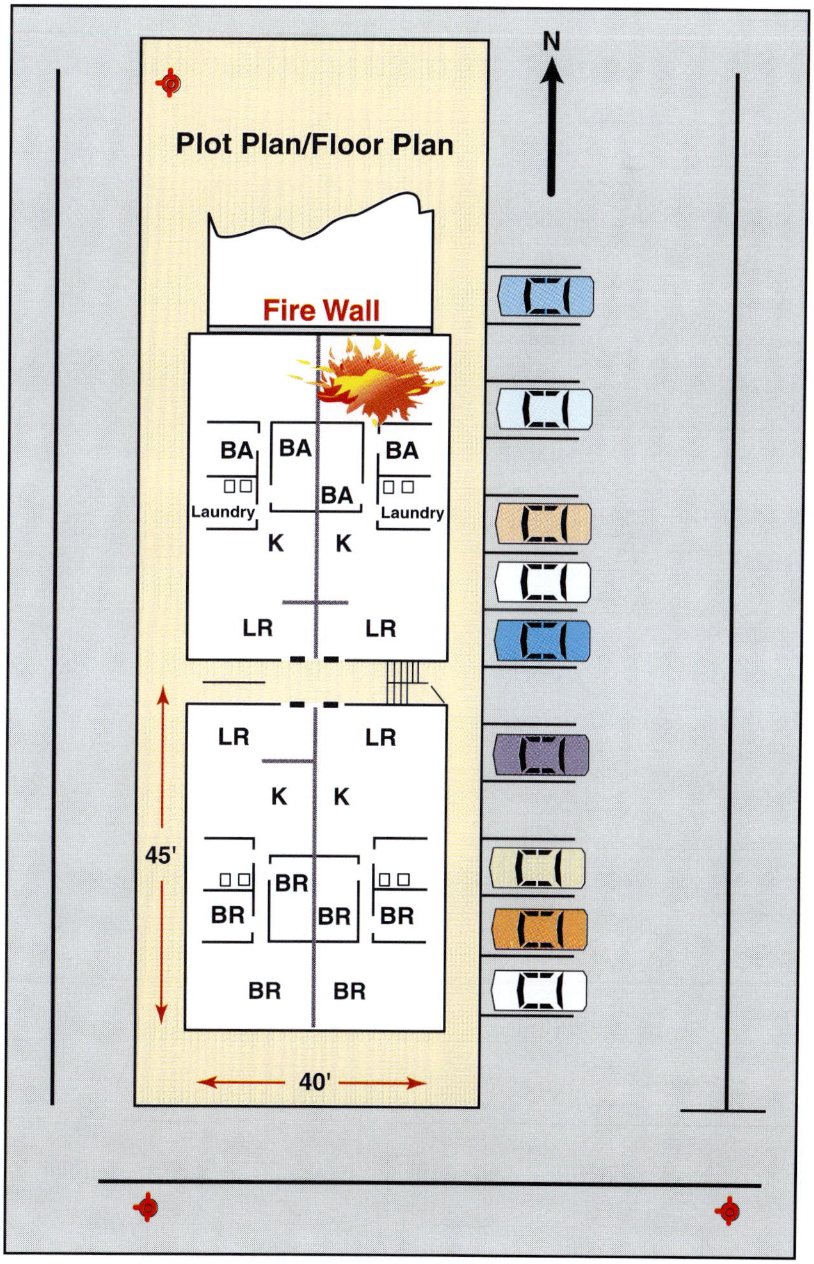

Activity 4.1

Part 1

Cues and Problems

From the scenario, plot plan, and your knowledge of garden apartments, list the cues that pertain to this fire situation. We'll start you off with a couple.

Cues:

Garden apartment – life hazard

Single entrance to vestibule and apartments

Problem Identification:

Occupants possibly trapped or overcome

Activity 4.1

Part 2

Strategy

Name the strategies that apply to the first-in company officer and chief at this incident. Use Lloyd Layman's RECEO VS from chapter 2. List strategies by priority of accomplishment.

Describe the tactic or tactics that you would use to accomplish each strategy.

Strategy #1 ____________________

Tactic 1 ____________________

Tactic 2 ____________________

Tactic 3 ____________________

Strategy #2 ____________________

Tactic 1 ____________________

Tactic 2 ____________________

Tactic 3 ____________________

Strategy #3 ____________________

Tactic 1 ____________________

Tactic 2 ____________________

Tactic 3 ____________________

Strategy #4 ____________________

Tactic 1 ____________________

Tactic 2 ____________________

Tactic 3 ____________________

Strategy #5 ______________________

Tactic 1 ____

Tactic 2 ____

Tactic 3 ____

Activity 4.1

Part 3

Tactical and Task Assignments

Using a two-alarm assignment of five Engines, two Ladders (Trucks), two Medics, and a Chief, describe each unit's assignments and duties. These assignments should correspond directly to the strategy and tactics in Part 2 of this exercise.

Engine 1: ______________________________

____

____

Engine 2: ______________________________

____

____

Engine 3: ______________________________

____

____

Truck 1: ______________________________

____

____

Engine 4: ______________________________

____

____

Engine 5: ______________________________

____

____

Truck 2: ______________________________

____

____

Medic 1: ____

____

____

Medic 2: ____

____

____

Chief 1: ____

____

____

Activity 4.1

Part 4

ICS Organization Chart

Based on your strategy and projected assignment of resources, complete an ICS organization chart for your operation.

Suggested Activity Answers

Activity 4.1
Part 1: Cues and Problem Identification

Cues:

Garden apartment—life hazard

Single entrance to vestibule and apartments

Smoke and flame from bedroom window on third floor

Problem identification:

Fire in one apartment—probably a bedroom

Fully charged apartment of origin

Possible occupants overcome in apartment of origin

Possible occupants overcome in other apartments on the third floor

Possible extension into attic space

Possible extension through fire wall into other section

Insufficient resources on hand for all the jobs to be done

Part 2: Strategy and Tactics
Strategy:

Rescue and primary search of the building

Exposure protection: check for extension in adjacent section

Confinement: check attic and other adjacent apartment

Extinguishment: put out the fire

Ventilation: remove smoke from third floor

Tactics:

Rescue strategy. Assign one company to search third floor and reinforce as necessary; search the second and first floors to make sure everyone is outside.

Exposure strategy. Assign one company to check for extension in the adjoining section (other side of fire wall).

Confinement strategy. One hose line into the apartment of origin; a second hose line brought to the third floor as backup or to extinguish any extension into attic or adjacent apartment; a third line may be required if the second line is aggressively used to fight the possible attic fire.

Extinguishment strategy. The hose line in the apartment of origin should be able to extinguish a room and contents fire.

Ventilation strategy. Horizontal ventilation of the third floor; if there is an attic fire of any significant magnitude, vertical ventilation may be necessary.

Part 3: Tactical and Task Assignments

Engine 1. Connect to a water supply, extend a 1¾-inch line, and attack the fire in the apartment bedroom. Call for a second alarm and two additional Medical Units.

Engine 2. Connect to a water supply and extend a 1¾-inch line to the third floor. The officer would be assigned Division/Sector 3 by the IC and given command of Engine 1's interior crew.

Engine 3. Designated as the Rapid Intervention Crew and stands by at the Command Post.

Truck 1. Would be assigned Rescue Group/Sector by the IC. This company would work in pairs to primary search the building and vent as they go. One team, before beginning the search, would immediately ventilate the fire room windows.

Engine 4. Would be assigned to check for extension in the section on the other side of the fire wall. The officer would be designated as Exposure Division/Sector D.

Engine 5. Would be assigned to assist the Rescue Group/Sector in primary search of the rest of the building.

Truck 2. Would be assigned as Vent Group/Sector to assist the engine companies with smoke removal and forcible entry into walls and ceilings.

District Chief 1. Assume IC on arrival and reassign the officer from Engine 1 to his crew working for the officer of Engine 2, who is Division/Sector 3.

Medic 1. Be assigned as Medical Group/Sector by the IC. They will establish an area to locate and treat victims and assess the situation for additional medical units.

Medic 2. Assigned to the Medical Group/Sector.

Part 4: ICS Organization Chart

Two Alarms (example of a correct ICS organization). Other organization charts may be correct. As long as your assignments have covered all of the necessary jobs on this incident, you are correct. You may want to have a departmental chief evaluate your work.

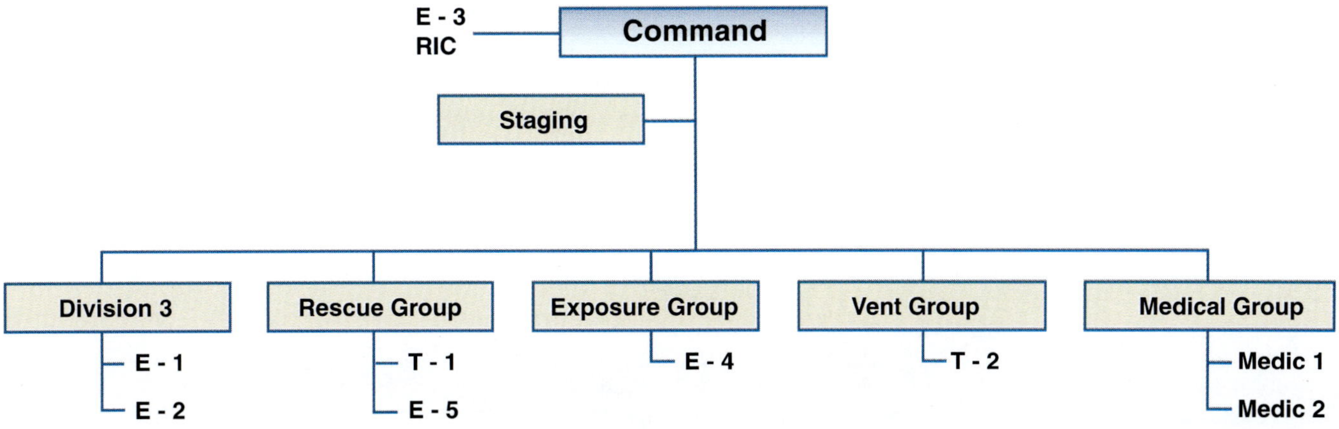

Chapter 5
Strip Shopping Centers

Strip Shopping Centers

Chapter 5

Introduction

A strip shopping center is a one- or two-story structure that contains two or more businesses or mercantile occupancies on the ground floor (Figure 5.1).

The construction for these shopping centers may vary among a number of types. Most are of ordinary construction with masonry walls and wood roof assemblies. Many of the more modern strip shopping centers are of noncombustible construction with masonry walls and steel bar-joist roof assemblies covered with metal decking. The top layer is a built-up roof of tar or tar and stone. Most are one story and are built on a concrete slab with no basement. However, many areas of the country have a number of strip shopping centers with two-stories and/or basements.

Figure 5.1 Typical one-story strip shopping center.

Figure 5.2 Typical two-story strip shopping center.

The second-floor occupancies of some strip shopping centers, especially the newer ones, may include office areas or apartments. Large (long) strip shopping centers may be separated by firewalls and may have anchor stores (Figure 5.2).

Occupancy-Specific Cues

Wall Construction

Typically, a strip shopping center's walls are constructed of masonry. The most common material is concrete block, but you may encounter poured concrete, tilt-up construction where the finished wall sections are set in place and anchored to the floor, to each other at the corners, and to the roof assembly. A relatively small number of strip malls use wood frame wall construction. Their walls are built of nominally sized wood (e.g., two-by-fours) similar to most private dwellings.

Another strip-mall construction method uses a skeletal steel frame to hold the roof assembly with nonbearing exterior walls. This construction type may be either noncombustible or ordinary, depending on the type of roof construction.

In more modern construction, firewalls or fire division walls may separate single or multiple occupancies. The attic space of these structures covers an individual occupancy. This reduces the probability and rapidity of fire spread, but it does not necessarily reduce the likelihood of collapse over a single involved occupancy. This is because collapse depends more on roof construction (Figure 5.3). Presently, most strip malls do not have firewalls or fire division walls. This ensures a common attic or cockloft over all the occupancies. Once fire reaches the common attic, fire spread often is very rapid and is destructive to the roof-assembly-bearing members.

Roof Assemblies

Strip shopping centers may have either flat or pitched roofs. The most commonly used flat-roof assemblies are wood beam and rafter with

Figure 5.3 The type of roof construction affects the likelihood of collapse.

sheathing, steel-bar joist, and parallel-chord wood truss and plywood I-beam. The most commonly used pitched-roof assemblies are ridgepole-and-rafter and pitched wood truss.

Wood Beam and Rafter with Sheathing. In this type of flat roof, beams are laid horizontally from the bearing walls and may be supported at several places between the bearing walls by lally columns. The beams, which are the main supports of the roof assembly, will carry the roof rafters. The roof rafters, two-by-six to two-by-twelve or greater solid lumber, are laid across the beams on centers ranging from 16 inches to 4 feet, depending on the rafter size. The rafters are covered with sheathing (one-inch thick lumber, plywood, or particle board), and then with a built-up rain covering. Generally found in older buildings, *this type of roof assembly has a very high resistance to collapse.* It normally burns through only at the involved fire area with very little weakening to the rest of the structure. Of course, in major fire situations where the entire attic space is well-involved, significant failure can result. If the risk of sending people under the roof assembly is high, then the risk of placing them on it is also high.

Steel Bar-Joist. These flat roofs use steel bar-joists suspended between the bearing walls. A steel column and beam system may act as a bearing member between bearing walls when the space being spanned is excessive for the bar-joist size. The bar-joists may be covered with wood sheathing and a rain covering as in Figure 5.4, or they may have Q-decking. Q-decking is a

Figure 5.4 Small strip shopping center.

sheathing made of corrugated sheet metal over which a rain covering is applied. The steel bar-joist construction method is used in both modern and older buildings. Steel bar-joist roof assemblies have been known to fail in as little as *nine minutes* after flame impingement. Failure is usually widespread over a large portion of the roof assembly.

Parallel-Chord Wood Truss and Plywood I-Beam. These types of flat roofs are found in modern construction. Born from the prefabricated and precut housing industry as a use for scrap lumber, the parallel-chord wood truss is a replacement for the solid wood rafter and the steel bar-joist in many applications. The plywood I-beam also has replaced the solid-wood rafter. The plywood I-beam is an assembly that has two-by-threes as top and bottom chords on a piece of ⅜-inch plywood. The wood rafters are covered with sheathing (plywood or particle board) and then with a built-up rain covering. Under fire conditions, both of these replacements have remarkably short failure rates. It is not unusual for both the parallel-chord wood truss and the plywood I-beam to fail after only *five minutes* of major fire involvement.

Ridgepole-and-Rafter. The ridge pole-and-rafter pitched roof creates a usable space in the attic area. The area that it covers is open and unobstructed. The ridgepole is located at the ridge, or peak of the roof. It is usually a two-by-six or greater board. Rafters are laid from the ridgepole to the top of the exterior bearing wall and nailed in place. The rafters are covered with sheathing (one-by-six boards are typical in older construction, plywood or particle board in more modern construction). The sheathing is covered with felt paper and shingles, tiles, or rolled roofing.

Pitched Wood Truss. Pitched trusses may be nailed at the juncture points of the webs and chords and other points where two separate pieces of lumber must be joined. Originally, wood pieces were simply toe-nailed together. Later, plywood was used as a gusset-plate material through which the nails were driven. This method provided added strength and faster assembly. Sheet metal gusset-plates also are used. The most recent technology is the gang-nailer. These are pieces of heavy sheet metal with ⅜-inch punched points protruding from one side. Every square inch of the gusset-plate has a number of these points. The gang-nailers are placed over the wood pieces that must be joined and then pressed into the wood. All of these truss assemblies have high failure rates under fire conditions. The gang-nailer may fail within *five minutes* of major fire involvement.

Rear Entry

Rear windows are limited or nonexistent in strip shopping centers. When windows are present, they typically are small, barred, and heavily screened (Figure 5.5). Rear door types include steel, steel-clad, or solid-wood. All are normally locked securely, even when the business is open. This makes rear

Figure 5.5 Bars and metal-clad doors are common barriers.

entry into these structures extremely time consuming when forcible entry must be accomplished, so most fire departments choose front entry even when it goes against the "burned to the unburned" philosophy.

WARNING

When people are known to be trapped between the fire and the rear of a strip shopping center occupancy, you must judge whether the front or rear offers the best chance for rescue. The built-in security features on rear entrances may present a situation that requires you to violate the tactical principal of "attacking from the unburned side."

Two-Story-Building Floor Assemblies

Beam and wood joist with sheathing floor assemblies similar to those for flat roofs often are found in older strip shopping center construction. Steel bar-joist floor assemblies may be used, and the second-floor surface may be a concrete deck. Parallel-chord wood truss or plywood I-beams are found in modern construction. These are covered with a sheathing of plywood or particle board and then carpet, wood flooring, or tile.

Comman Problems

Roof or Ceiling Assembly Construction Type

A prefire plan *must* identify the type of roof construction in a one- or two-story strip shopping center ahead of time. In a two-story strip center, the second floor assembly also needs to be identified. You cannot make the risk-benefit analysis of whether to place personnel over or under an assembly unless you know the type of construction materials used. After arrival at the fire scene is a poor time to gather roof and floor construction information. *If you do not know the type of roof construction, assume that it is lightweight and subject to early collapse once fire has gotten a foothold on the assembly.* Take the time to determine the roof and floor construction materials by going to an adjacent occupancy and pulling some ceiling.

Note: When initially pulling ceilings, crews should be positioned near doorways/walls to provide some protection should the ceiling collapse en masse.

Fire Involvement Considerations

Has the fire extended to the roof assembly area? Has it extended to the first-floor ceiling assembly of a two-story structure? What are the signs of probable roof/ceiling assembly involvement?

- Major fire involvement in one occupancy
- Fire involvement in two or more occupancies
- Hot, rising smoke from attic vents
- Fire venting through the roof

Is the fire likely to spread to the roof assembly soon? What are the signs that the fire will likely spread to the roof or ceiling assembly?

- One or more well-involved single occupancies
- Fire is not being controlled in a timely manner due to insufficient fire flow into the area of involvement.

Structural Deterioration and Collapse

In one-story structures, collapse potential is related directly to roof construction type. Any type of truss construction can be expected to collapse early once it has become well involved in fire (Figure 5.6). Collapse can come as soon as five minutes after involvement. In two-story structures, collapse potential is related to both floor and roof assembly construction. Older buildings may have structural deficiencies even before the fire. Loads on the roof, such as HVAC systems, may accelerate collapse.

Figure 5.6 A well involved common attic may result in a rapid collapse of the roof assembly.

Cue-Based Predictions

By applying the occupancy-specific cues in a strip shopping center fire situation, we can make a number of predictions and assign Groups/Sectors accordingly.

Rescue Group/Sector

The life hazard is likely to be minor on the ground floor where the mercantile occupancies are located. United States Fire Administration statistics do not indicate a fire death problem at strip shopping centers; however, this does not preclude the need for a primary search. The life hazard in any residential occupancies above a strip shopping center is very high. You can expect heat, smoke, and toxic gases on the second floor when the first-floor occupancy is well-involved.

Exposure Group/Sector

Fire often occurs after closing time and gains considerable headway. This allows time for extension to floors and attic spaces above. Fire travel in these floor and attic spaces can spread the fire to other occupancies.

Access to the rear of the structure may be very time consuming due to security devices. A delayed access will mean that the fire will increase in intensity and size. It may be possible to effect a rescue or attack a fire by breaching a wall from the adjacent occupancy.

Incident Management Cues

Primary Search and Ventilation

Conducting a primary search in a strip shopping center is just as important as in any other building. The IC should assign at least one company to this task. Ventilation efforts will speed the search crews' actions. The faster vertical and horizontal ventilation is completed, the faster crews can complete the primary search and be reassigned to fire-fighting duties.

Confinement and Vertical Ventilation

Vertical ventilation is critical for interior crews to stop the lateral spread of fire in a common attic area. The common attic space (cockloft) allows entering smoke and heat to travel horizontally toward the end of the building. This travel, called mushrooming, will obviously cause the fire to spread to additional occupancies adjacent to the original fire occupancy. Vertical ventilation stops the downward and horizontal movement of smoke and heat into the operating area and adjacent stores. Crews can see farther and work faster when visibility is above zero.

Strategy and tactics

Occupancy-Specific Strategies

Based on the occupancy, few cues will indicate rescue at a strip shopping center fire. Rescue needs generally will be minimal on the first floor but greater on the second floor if it is residential.

Fire control cues for strip shopping centers include building construction, floor construction, and roof construction. A common attic space shouts the need for extension control. The location and extent of the fire will indicate the resources needed for the various tactics that must be performed.

Ventilation cues include smoke in the fire occupancy and adjacent occupancies. Smoke coming from roof vents may indicate involvement of the attic space. Immediate vertical ventilation will be required. Horizontal ventilation also must occur early in order to permit operating forces to work effectively in the exposed areas.

Fires Not Involving the Roof Assembly—One-Story

- Begin an aggressive interior attack on the fire with horizontal ventilation.
- Closely follow the initial attack by assigning personnel to check for extension.
- Check ceiling space and exposures.

Fires Not Involving the Roof Assembly—Two-Story

1. Two-story buildings with a fire on first floor.

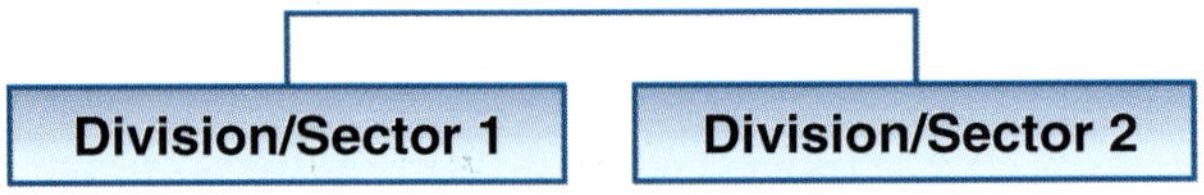

- Check the adjoining occupancies;

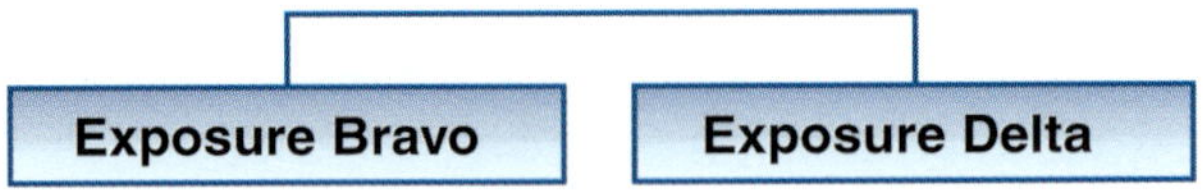

- first floor;

- ceiling space; and
- occupancy directly above the fire.

2. Two-story buildings with a fire on second floor.
   - Check exposures on both sides;

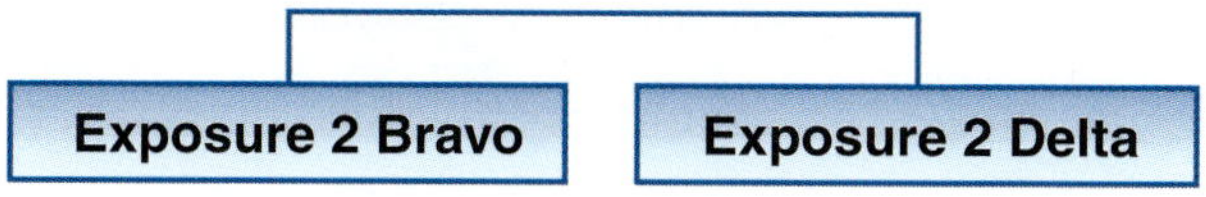

   - attic space; and
   - occupancy below the fire.

Fire in a Roof or Ceiling Assembly over a Single Occupancy Separated from Other Occupancies by Firewalls or Division Walls

- Roof or ceiling construction must be known before an interior attack or vertical ventilation is initiated. If you do not know the type of construction, *assume that it is lightweight.* Then check an adjoining occupancy for the information by pulling some ceiling.
- You must assess risk/benefit before placing personnel on or under a roof or floor assembly of truss construction.
- Recommended tactics when the risk is acceptable:

— Vertically ventilate the roof over the fire area.

— Horizontal ventilation front and rear.

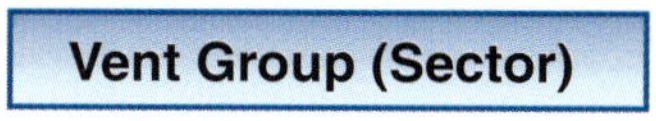

— DO NOT put a ladder pipe or other exterior stream into any ventilation hole.

— Interior fire attack—apply large quantities of water on the fire.

— Operate from an area between the fire and an exit. Pull ceiling from the doorway and begin extinguishment. DO NOT advance into the room with fire over your head and allow it also to be behind you.

— Check interior exposures and attic areas for extension.

Well-Involved Fire in a Roof or Ceiling Assembly That Serves More than One Occupancy

- Roof or ceiling construction must be known before an interior attack or vertical ventilation is initiated. If you do not know the type of construction, *assume that it is lightweight.* Then check an adjoining occupancy for the information by pulling some ceiling.

- You must assess risk/benefit placing personnel on or under a roof or floor assembly of truss construction.
- Recommended tactics when risk is acceptable:
  - — Perform vertical ventilation over involved occupancy.
  - — Perform horizontal ventilation front and rear.
  - — DO NOT put a ladder pipe into any ventilation hole.
  - — Check interior exposures for extension.
  - — Start at least two occupancies away on each side to make sure you are ahead of the fire.
  - — Pull ceilings to check exposed roof or floor areas for extension and provide a path for water to reach the fire.
  - — Control horizontal extension by applying large quantities of water to keep the fire from extending.
- Recommended tactics when the risk is unacceptable:
  - — Launch exterior fire attack using large-caliber streams.
  - — Perform horizontal ventilation.
  - — Check interior exposures for extension.
  - — Check exposed attic areas for extension.

Activity 5.1
Strip Shopping Center Fire

Read the scenario of a fire situation in a strip shopping center and examine the accompanying photograph and plot/floor plan. Then assume the role of the first-in officer at the incident and complete this activity's four parts. First, determine the cues and the problems that are presented in the scenario and the plot/floor plan. Second, determine the strategy and tactics for the incident. Third, specifically identify the tactics or tasks that will be performed by the resources you call to the scene. And fourth, draw an Incident Command System organization chart for the operation.

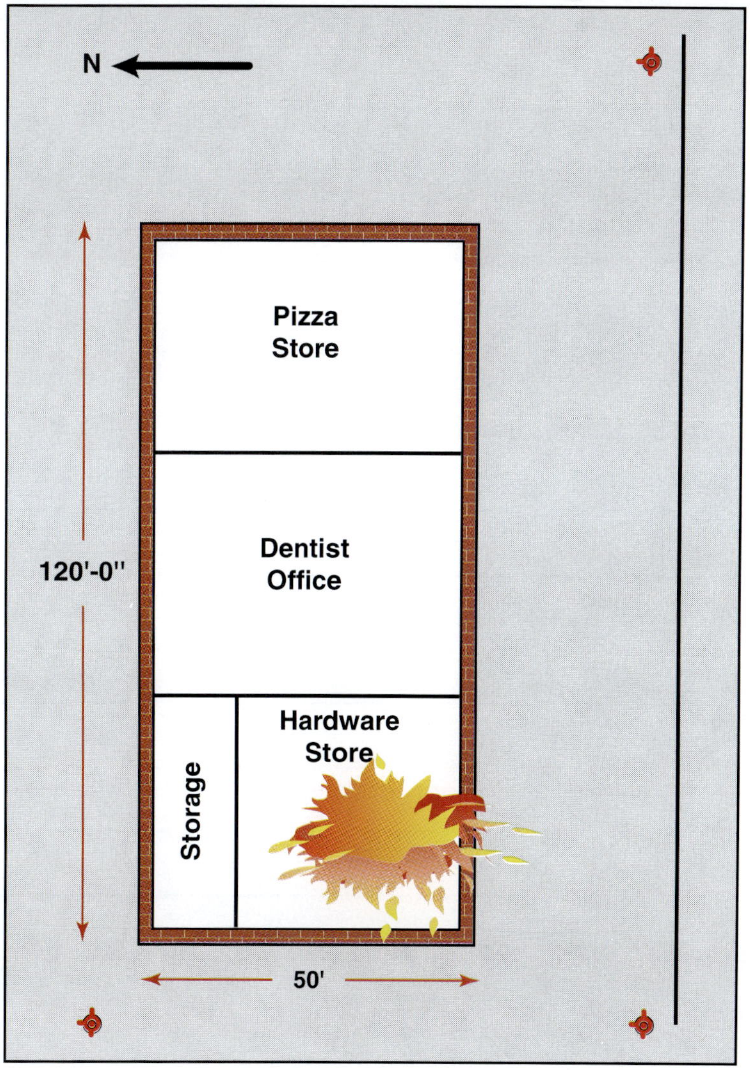

Scenario

You are responding to a strip shopping center fire as a Chief Officer. Three Engines, a Ladder Company, and a Medic Unit are on the assignment with you. Your prefire plan indicates that the structure has a common attic over the three occupancies and that it is of ordinary construction with a beam and rafter roof assembly. As you approach the scene, you see heavy smoke and flame exiting the front of the hardware store at one end of the strip shopping center. It is 0600 hours. There is heavy smoke in the dentist's office and light smoke in the pizza store.

Activity 5.1 (cont.)

Part 1
Cues and Problems

From the scenario, plot/floor plan, and your knowledge of strip shopping centers, list the cues that pertain to this fire situation. We'll start you off with a couple.

Cues:

Strip shopping center—low life hazard

Rear entrances locked securely

Problem Identification:

Fire in one occupancy at this time

Activity 5.1 (cont.)

Part 2
Strategy and Tactics

Name the strategies that apply to the first-in company officer and chief at this incident. Use Lloyd Layman's RECEO VS from chapter 2. List strategies by priority of accomplishment.

______________________________ ______________________________

______________________________ ______________________________

______________________________ ______________________________

______________________________ ______________________________

______________________________ ______________________________

______________________________ ______________________________

______________________________ ______________________________

Describe the tactic or tactics that you would use to accomplish each strategy.

Strategy #1 ______________________________

Tactic 1 ______________________________

Tactic 2 ______________________________

Tactic 3 ______________________________

Strategy #2 ______________________________

Tactic 1 ______________________________

Tactic 2 ______________________________

Tactic 3 ______________________________

Strategy #3 ______________________________

Tactic 1 ______________________________

Tactic 2 ______________________________

Tactic 3 ______________________________

Strategy #4 ______________________________

Tactic 1 ______________________________

Tactic 2 ______________________________

Tactic 3 ______________________________

Strategy #5 ______________________________

Tactic 1 ____

Tactic 2 ____

Tactic 3 ____

Activity 5.1 (cont.)

Part 3
Tactical and Task Assignments

Using a two-alarm assignment of five Engines, two Ladders (Trucks), two Medics, and a Chief, describe each unit's assignments and duties. These assignments should correspond directly to the strategy and tactics in Part 2 of this exercise.

Engine 1: ____

____

____

Engine 2: ____

____

____

Engine 3: ____

____

____

Truck 1: ____

____

____

Engine 4: ____

____

____

Engine 5: ____

____

____

Truck 2: ____

____

____

Medic 1: ____

____

____

Medic 2: ____

____

____

Chief 1: ____

____

____

Activity 5.1

Part 4
ICS Organization Chart

Based on your assignments, complete an ICS organization chart for your operation.

Suggested Activity Answers

Activity 5.1

Part 1: Cues and Problem Identification

Cues:

Single-story strip shopping center—low life hazard

Major fire in hardware store—involvement toward the front of the store

Fire involvement is about 50 percent—likelihood of fire in the attic space

Problem identification:

Fire in one occupancy

Fire possible in common attic space

Heavy smoke in hardware store

Heavy smoke in dentist's office

Light smoke in pizza store

Possible hazmats in hardware store

Bottled gas in dentist's office

Part 2: Strategy and Tactics

Strategy:

Primary search (rescue) in dentist's office and pizza store.

Confinement—contain the fire to the hardware store.

Extinguishment—put out the fire.

Ventilation—both horizontal and vertical.

Tactics:

Rescue (primary search) strategy. Assign one company officer the responsibility to do a quick search with his/her company.

Confinement strategy. Enter the dentist's office with two 1 3/4-inch lines and pull ceiling to stop fire in the attic space.

Extinguishment strategy. Place a master stream on the hardware store from the front and additional hand lines front and rear as needed.

Ventilation strategy. Vertically ventilate over the rear of the hardware store to relieve pressure in the attic space. Horizontally ventilate all three stores, front and rear, by breaking glass and forcing doors.

Part 3: Tactical and Task Assignments

Engine 1. Connect to a water supply and park directly in front of the hardware store. Supply one 1¾-inch hand line to the dentist's office for the two firefighters to operate. The pump operator will set up a master stream on the main body of fire as soon as the supply line and hand line are charged. Call for a second alarm. The company officer assumes Command.

Engine 2. Connect to a water supply and proceed to the front of the building. The officer would be assigned as Exposure Division/Sector D by the IC and directed to take charge of Engine 1's crew, which is operating in Exposure Division/Sector D. Engine 2 would enter with a second 1¾-inch hand line and proceed to do a primary search and to pull ceiling and stop extension in the attic space. The company officer would request one additional company to assist with the operation.

Engine 3. Designated as the Rapid Intervention Crew (RIC) and stands by at the Command Post after doing a quick search of the pizza store.

Truck 1. Would be assigned as the Vent Group/Sector and directed to vertically ventilate the hardware store at the rear. This company would also be assigned horizontal ventilation. The company officer would request one additional company to carry out the horizontal ventilation from the front and the forcible entry from the rear.

Engine 4. Would be assigned to Exposure Division/Sector D as a third company.

Engine 5. The company officer would be assigned to extinguish the fire in the hardware store and would be assigned as Division/Sector A. Engine 1's pump operator would work under this officer since there is a master stream on the fire from Engine 1.

Truck 2. Would be assigned to Vent Group/Sector to force entry in the rear of the occupancies.

District Chief 1. Assume IC on arrival and reassign the officer from Engine 1 to his crew working for the officer of Engine 2, who is Exposure Division/Sector D.

Medic 1. Directed to stand by at Command Post to assist IC with management needs (radio, ICS chart, etc.).

Medic 2. Directed to remain at Staging.

Part 4: ICS Organization Chart

Two Alarms (example of a correct ICS organization). Other organization charts may be correct. As long as your assignments have covered all the jobs that have to be done on this incident, you are correct. You may want to have a departmental chief evaluate your work.

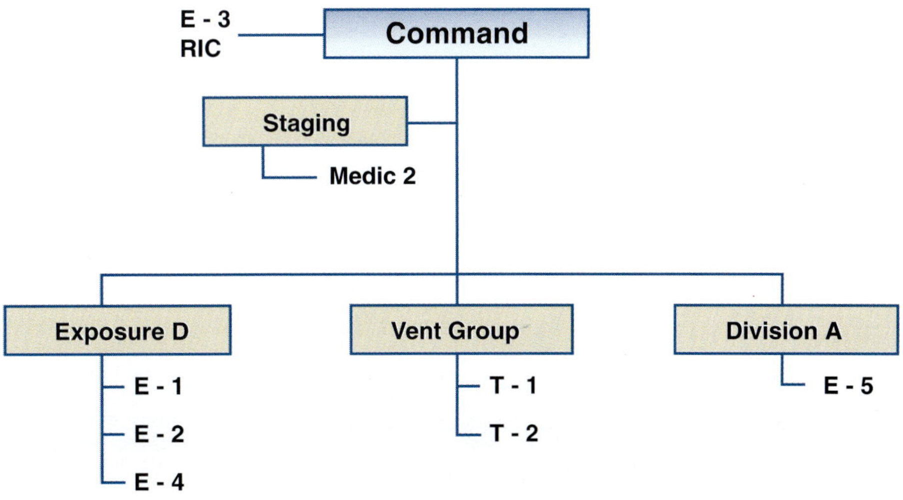

Chapter 6
Vacant Commercial Buildings

Vacant Commercial Buildings

Chapter 6

Introduction

A vacant commercial building usually is defined as a building or structure that was used for manufacturing, warehousing, or sales and currently is not occupied.

Buildings that fit this classification may be of various design and construction types and may be either single story or multiple story (Figure 6.1). They can range in size from small to very large.

These types of buildings can be vacant for many reasons. Often the business that occupied the structure has failed, or the structure is scheduled for demolition because of its age or condition or to allow construction of another building. Others may be vacant while waiting for new owners/lessors to occupy the premises or for renovations to be completed.

Figure 6.1. Vacant occupancy at the end of a shopping center.

Occupancy-Specific Cues

Special Construction Factors

The type of construction found in vacant commercial buildings will vary greatly. The range of construction types will run the gamut from noncombustible to fire resistive to ordinary and, in some cases, even wood frame, depending on the design and age of the building and the use for which it was intended originally. Mixed construction types are common in vacant commercial buildings, especially older buildings that have been expanded, remodeled, or modified for specific types of occupancies or uses.

Vacant commercial buildings, often by their very nature, contain construction deficiencies that must be considered during fire fighting. Expansion or modification of the structure may have created conditions that will affect fire spread and the general stability of the building.

Those vacant commercial buildings that pose the most serious problems under fire conditions are typically of old-style construction, have minimal fire resistiveness, lack built-in fire protection, and have the potential for rapid fire spread and early collapse. These buildings are generally of ordinary construction (masonry walls with wooden floor and roof assemblies).

Wall Construction

Wall construction in vacant commercial buildings also can be of many different types. Masonry walls, which may be concrete block or brick and which may be reinforced, are quite often found in these types of structures. On the West Coast, many of these buildings were constructed with tilt-up systems and will have poured concrete walls. Metal walls or combination metal-masonry walls also are typical of vacant commercial structures. Some buildings, especially multistory structures, may be of steel frame with curtain walls.

When wood-frame walls are found in vacant commercial structures, their construction is usually of the platform type, which limits the unrestricted vertical travel of fire within the walls.

In some cases, firewalls or fire division walls may divide single occupancies or may separate multiple occupancies within a single structure. This is a best-case scenario for the fire service, however, and it is found infrequently.

Always consider the potential for wall collapse in vacant commercial structures. The failure of other structural members during a fire can trigger a wall collapse. The failure of the roof assembly is quite often the cause for wall failure. The potential for wall failure should be monitored throughout the fire. The Safety Officer should monitor the structure for possible deterioration during fire operations and apprise the Incident Commander of his findings and predictions about roof or wall failure.

Roof Assemblies

Roofs on vacant commercial structures will vary depending on the building's age, design, and construction type. Flat roofs that consist of wood beams and rafters and wood sheathing are typical of older buildings. Many newer and some older buildings feature steel bar-joist roofs that can fail very quickly under severe fire conditions. Steel bar-joist roofs have been known to fail in as little as nine minutes after significant fire involvement.

Parallel-chord wood truss and wooden I-beam roofs are typical of modern construction designs and can be found on many newer buildings that have been vacated. Panelized roofs, which are used on many tilt-up buildings, use girder and beam construction with minimal support of the plywood roof decking. These roofs are extremely dangerous to be on during fires.

The two general types of raised-roof designs on vacant commercial structures are pitched roofs and arched roofs. Pitched roofs can be of two different designs. Ridgepole-and-rafter roofs commonly are found on older, smaller buildings. Wood truss pitched roofs also are common. The wood truss roofs may be nailed (typical in older construction) or gusset plate (typical of more modern truss construction). Arched roofs are typical of large-area buildings and may be of bridged-truss or bow-string design. Bridged-truss roofs are supported by rafters and are considered to be relatively safe under fire conditions. Bow-string roofs, on the other hand, have a reputation for collapse under fire conditions, especially if the bottom chord is loaded with additional weight, such as stored materials.

Fire Loading and Fire Spread

Fire loading will vary greatly in vacant commercial structures, based on general building conditions. Buildings of this type that have been cleaned out and properly secured usually have minimal fire loading. Buildings where some debris and combustible fixtures or furnishings remain are considered to have medium fire loading. Maximum fire loading is present when the building contains large amounts of combustible materials and is unsecured.

Delayed alarms are common with major fires in vacant commercial structures. Obviously, when a building is not occupied and the fire is not reported promptly, special problems are created for responding firefighters. Among these are the lack of any reliable indication as to how long the fire has been burning. This creates the potential for possible damage to structural bearing members within the building and the probability of *flashover* or *back draft*. These two conditions, deterioration of bearing members and flashover or back draft, are very likely when the alarm is delayed.

Fire spread in vacant commercial structures may be rapid for a variety of reasons. Large amounts of combustibles within the building and the fact that the building may be wide open can create very fast-moving fires. Unrestricted paths for the fire to travel through open stair shafts or building deficiencies also will cause the fire to spread very rapidly.

The incidence of *incendiary fires* in vacant commercial properties is high. Vagrants and homeless people often inhabit vacant commercial buildings. They build fires in the buildings and end up burning the buildings down. In other incendiary cases the owners are "selling" the building to the insurance company. Many times, incendiary fires will have multiple areas of origin, which causes rapid fire development and spread. Because the potential for incendiarism is quite high, caution is required during initial entry and during fire attack operations.

Floor Assemblies

Floors in vacant commercial structures may be beam and joist with wood decking, which is typical in older buildings, or concrete, which is commonly found in newer buildings. Many multistory buildings may have concrete first floors with beam and joist floors above. Trussed joist floor assemblies, such as parallel-chord wood truss, also are common on some newer buildings. *The possibility of floor voids in vacated buildings requires that extra caution be taken during interior operations.* Loss of floor girders or floor beams and joists may trigger wall collapse in some types of construction. This is especially true in older style buildings such as heavy timber, ordinary construction, and those with unreinforced masonry walls.

Common Problems

Roof, Wall, or Floor Assembly Construction Type

Roof, wall, and floor construction of the building can be critical during a fire situation (Figure 6.2). These should be identified during preplanning.

Figure 6.2. Vacant occupancy in the center of a strip shopping center.

Time and conditions usually will not permit detailed determination of these elements on arrival at the incident scene. If you cannot determine the construction of key elements of the building, then assume it is lightweight construction and act accordingly; allow no one on the roof or under the roof if the roof assembly is well-involved.

Fire Involvement Considerations

Fire involvement on arrival must be a primary size-up consideration. Determine the location of the fire in the building and the amount of fire as quickly as possible. Also check extension of the fire from the original area of involvement. Signs of potential fire spread are the volume of fire in the building, the horizontal and vertical travel paths for the fire to spread, and the fire's location in terms of the direction from which it can be attacked. If the building is obviously lost, do not endanger your personnel with offensive tactical operations.

Immediately check extension of the fire through the ceiling into the attic area. Signs of probable roof or attic involvement are major fire involvement of the interior of the building or fire venting through the roof. Hot smoke rising forcefully from attic vents or heat discoloration of the roof covering also indicate the fire may be burning in the ceiling or attic area. You must also determine whether the fire is likely to affect any primary structural components, such as the roof, wall, or floor assemblies.

Structural Deterioration and Collapse

Consider structural deficiencies in terms of overall safety and the method of fire attack. General decay or structural modifications will play a major role in fire development and spread. This also is true if the building is partially demolished and any fire protection features built into the building have been removed or rendered ineffective. A building whose value after the fire will be near zero calls for the defensive fire attack mode.

Structural deterioration of vacant commercial buildings may be rapid due to the amount of fire present on arrival and the uncontrolled spread of fire before it can be contained. The building's construction type as well as its general condition also will affect possible failure of structural components.

In these types of buildings, the roof generally will be the first structural element to fail. The type of roof construction is the key, with trussed roofs being the most susceptible to early failure. Anticipate roof collapse if the fire has been burning for a prolonged period or if a large volume of fire has extended to the roof or attic area.

Wall collapse may be gradual or sudden, depending on construction type. Often the shift in walls from normal vertical position is noticeable as the fire progresses, and this condition should be monitored. Wall collapse can be tied closely to the type of roof construction. The loss of the roof usually is preliminary to wall failure, as the collapse of the roof typically will affect the walls' horizontal and vertical stability. In some cases, roof and wall failure may occur at the same time.

The direction of wall collapse, either inward or outward, cannot be predicted. The loss of roof or floor structural members may push walls out or pull them in. As a general rule, walls in heavy timber or ordinary (masonry and wood-joist) construction will fall outward, but don't bet the lives of your personnel on this axiom. The rule of thumb is that walls will fall outward to a distance of one and one-half times their height; this should be considered the *minimum* safe distance, as the height of the wall and the impact of its parts with the ground may propel them a greater distance.

The collapse of floors may result from fire burning on the floor itself or on the floor below. This is especially true of truss-type floors. Floors may also collapse due to walls pulling away from the floor assemblies. This is common in ordinary construction, where the walls often bow out during fire conditions.

If the collapse of a roof, wall, or floor is indicated or anticipated, remove firefighters from inside the building, from the roof of the building, and from within a safety zone on the outside perimeter of the building. Remember, if a building is not safe to be in or on top of, it is not safe to be around.

Exposures

Exposures in vacant commercial structures must be a primary consideration in developing an initial strategy and determining tactical objectives (Figure 6.3). In vacant building fires, exposures will fall into two basic categories: connected exposures and unconnected exposures. The type of exposure generally will affect the overall strategy.

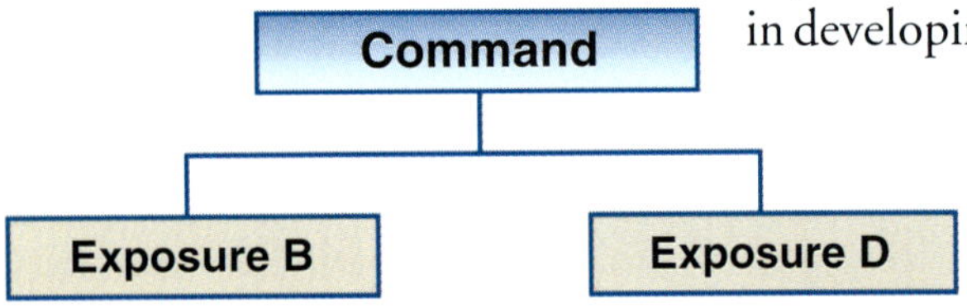

Figure 6.3. Vacant occupancy with a severe exposure problem.

Connected Exposures. Connected exposures may be difficult to protect due to their proximity to the fire area. Normally they will require you to place protective lines inside the exposure. Fire conditions and fire behavior may make interior protection difficult and extremely dangerous. This is especially true if there are no or only minimal protection features between the fire occupancy and the exposure. Lack of fire walls, fire division walls, and rated fire doors, coupled with common attics, are examples of elements that make connected exposures difficult to protect.

Unconnected Exposures. Unconnected exposures usually are much easier to protect than connected exposures. The space between the fire building and the exposure buffers the radiated heat and provides space to position protective hose streams. Protecting these types of exposures normally does not require interior operations and frequently can be accomplished with master streams placed at key points on the exterior of the exposure. The master streams may be used to knock down visible fire on the original fire building and to wet down the exterior of the exposed building.

Access Points

Access points into vacant commercial buildings are important. The ability to reach the fire area will be critical in terms of time required and the direction from which the fire is attacked. Because the access points' locations can play an important role in fighting the fire, they should be identified on preplans.

Entry into well secured vacant commercial structures may be difficult. Fire attack teams will require support assistance in order to enter the building. If entry is delayed because access points are well secured, the fire may grow considerably and be more difficult to control. Effective truck company operations are a necessity. Even if you have no ladder trucks, you must assign some other crew(s) to do truck work.

If possible, select access points that allow a fire attack from a direction that will not drive it into uninvolved areas of the structure. If this is not possible, then place additional lines in the structure to provide for interior exposure protection by holding the fire to its original location.

Cue-Based Predictions

When the fire is of significant proportions, the defensive mode will be applicable from the arrival of the first apparatus. First, ensure proper apparatus placement and then protect exposures and establish a high-volume water supply.

View the structural integrity of vacant commercials pessimistically when they are involved in fire. Fire spread to bearing members with deteriorated assemblies can cause walls, roofs, and floors to collapse suddenly.

Suspect arson at any fire in a vacant commercial structure, and develop your Incident Action Plan accordingly. Also keep in mind that master streams and large hand lines often are required to knock down these fires.

Incident Management Cues

Fires in vacant commercial structures often require multiple tactical operations. Interior attack, exposure protection, and ventilation are just a few of the activities that must be coordinated during these incidents. Failure to coordinate tactical operations can result in very ineffective fire control operations, but more importantly, can cause serious injury to fire control personnel. Command Officers at all levels must ensure that tactical operations are performed in a timely manner and do not conflict with each other. Opposing hose lines, mixing offensive and defensive tactics, and premature or delayed ventilation are just a few examples of activities that can cause problems if tactical operations are not controlled and coordinated.

Strategy and Tactics

You must perform a good size-up before making any tactical decisions to fight a fire in a vacant commercial structure. Some critical factors that you should consider during size-up are:

- Construction type
- Possible structural deficiencies
- Potential structural failures
- Location of fire in the building
- Amount of fire
- Predicted fire behavior
- Resource needs

Acting before thinking and planning at the scene of a vacant commercial fire can place firefighters in serious danger and hinder control effectiveness. A few moments spent making a proper size-up can prevent many of these problems. Size-up leads to the proper identification of all problems at the incident. Only when these problems are known should operations begin.

Strategy and tactics in these types of situations must be based on a risk/benefit analysis. The first questions that you must answer are, "Are there any life safety considerations?" and "Is this building worth saving at the risk of firefighter's lives?"

Your decision as to what type of attack mode you will use should take into account the commercial value of the building. What will be left of value after the fire in the building is extinguished? If most of the structure will be damaged or destroyed to the point that it will no longer be usable, then the strategy and tactics should be based on minimizing any safety concerns for firefighters. If the building is scheduled for demolition, which is the case with many vacant commercial structures, then the risks involved in trying to save the building far outweigh the benefit to be gained, and this should be the governing factor in deciding what action to take. Defensive tactical operations apply in this scenario.

Attack Mode

When you develop strategy and assign tactical objectives, exposures should be a primary concern. When multiple exposure problems exist, they should be handled in order of their priority. Occupied exposures and exposures connected to the fire building are of the most immediate concern and should be dealt with first. Exposures that are not connected or not occupied pose less of a threat and are lower in priority.

Your decision on the type of fire attack to use must be based on fire conditions and available resources. The attack mode may be either offensive or defensive and may change according to fire conditions and resource availability.

An offensive attack in these situations would involve an attempt to extinguish the fire and save the building. Current conditions must indicate that it is safe to put people inside the building to extinguish the fire or on top of the building to perform ventilation operations. If you are contemplating an offensive attack, the risks associated with such an operation must be acceptable. If they are not, employ a defensive attack. When you use an offensive attack, you must monitor scene conditions such as fire location, fire size, and structural stability constantly. Remember that fire conditions can alter rapidly and may require rapid changes in strategy and tactics.

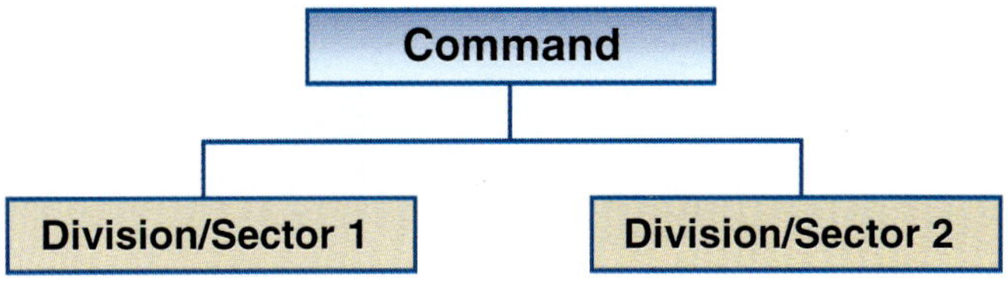

When mounting an offensive attack for anything other than a small fire, use larger diameter (2½-inch) hand lines. While smaller hand lines may be sufficient for other types of structure fires, larger diameter hand lines provide an increased margin of safety. They also provide greater reach and more extinguishing capability, which fires in vacant commercial buildings sometimes require.

When you use interior attack teams to fight fires in vacant commercial buildings, assign backup teams with charged hose lines to provide protection. Officers supervising interior operations must monitor the location and status of personnel inside the building constantly and notify them of any changes that may affect their safety.

A defensive attack on a fire in a vacant commercial building may take many forms. It may be an attempt to extinguish the fire in the building without actually placing personnel and hose lines into the structure. It may consist of placing hose streams from hand lines or heavy stream appliances in key positions to protect exposures. *Exercise caution when mixing handheld lines and heavy stream equipment during defensive operations.* Failure to do so can subject firefighters to unnecessary danger from hose streams and pushing of fire by master streams.

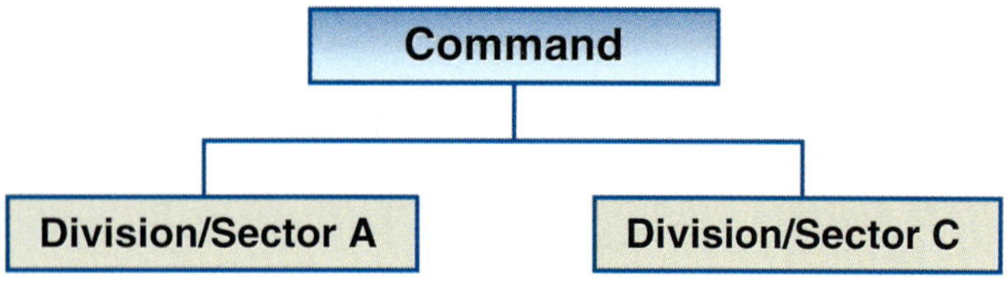

If master stream appliances might be used in defensive operations, consider initial placement of apparatus on arrival at the incident scene. Attempting to reposition apparatus for defensive operations when a change in attack mode is required can be extremely difficult. A little thought about where apparatus are located initially can make this process much easier.

In many cases, the initial strategy may be a defensive operation that attempts to hold the fire in check or to protect exposures pending the arrival of additional resources. When the required resources arrive on-scene, a switch to an offensive mode can often be made. Conversely, changes in fire conditions may require a switch from an offensive attack to a defensive attack. When this occurs, it is critical that everyone at the scene gets the message that the attack mode is changing and acts accordingly. This is the time when the IC and Division, Group, and Sector Supervisors must ensure that personnel stay well back from the structure walls.

It is also critical to employ a common signal that everyone at the scene understands to mean conditions indicate it is no longer safe to be inside or on top of the building. This often is a specific air-horn signal or a special dispatcher-generated wail over the radio system followed by a radio message to evacuate the building as soon as possible. This alarm signal can take many other forms, but when sounded it should alert everyone to leave the building and the immediate perimeter area immediately.

Activity 6.1
Vacant Commercial Buildings

Read each of the brief scenarios and then answer the questions that follow.

Scenario 1

A 90' × 100', one-story vacant commercial structure of noncombustible construction has a 50-percent involvement. The roof assembly is steel bar-joist with metal decking. There obviously was a delayed alarm.

What problems are created, and what tactical solutions and Incident Command System (ICS) solutions must be applied?

Problem A: ______________________________

____

____

Tactical Solution: ______________________________

____

____

ICS Solution: ______________________________

____

____

Problem B: ______________________________

____

____

Tactical Solution: ______________________________

____

____

ICS Solution: ______________________________

____

____

Problem C: ______________________________

____

____

Tactical Solution: ______________________________

____

____

ICS Solution: ______________________________

____

____

Scenario 2

A three-story, 80' × 80' vacant commercial of ordinary construction, circa 1935, has a 20- to 25-percent involvement on the second floor. The building has been vandalized, and all the copper tubing and electrical wiring have been removed from the walls and floors. The stairs are deteriorated or missing, and most of the windows are broken.

What orders would you give pertaining to the safety of fire fighting operations?

____

____

____

Scenario 3

A 60' × 80', one-story vacant commercial with a bowstring truss roof assembly has a 40- to 50-percent involvement. The initial Incident Commander (IC), Lieutenant from Engine 1, has directed Truck 2 to perform roof ventilation as well as horizontal ventilation. The IC also has directed two crews to initiate hand line operations into the structure. As the first arriving chief officer, what orders, if any, would you give to support or change the present operation?

____

____

____

____

____

____

Scenario 4

A 75' × 100', four-story vacant commercial structure of ordinary construction is involved in fire. Initial conditions indicate that the second floor is involved. Initial operations were ventilation and interior fire attack. Two alarms are operating at the scene. After approximately 15 minutes of fire attack, the fire has extended to the third floor, and now that floor is well involved. The fire fighting forces on the second floor have made partial headway in stopping the spread of fire there. The third-alarm companies have arrived at staging and are ready for assignment. As the IC, what, if any, decisions would you make at this time?

Activity 6.1 Answers

Scenario 1:

Problem A: The roof assembly will fail extremely fast.

Tactical Solution: Roof operations are not applicable in this situation. Apply master streams from the exterior. Allow no operation on or under the roof.

ICS Solution: Division (Sectors) A, B, C, D around the outside.

Scenario 2:

Due to the probable rapid spread of fire through the walls and floors and up the open stairs, operations should be conducted from the exterior and preparations made for full defensive operations.

Scenario 3:

Crews should be removed immediately from the roof and from any operations under the roof assembly. A defensive, heavy stream attack should be established from defensive positions.

Scenario 4:

When two floors of an ordinary constructed building are well involved for 20 minutes, the building is likely to collapse soon. Crews should be removed immediately, and the operation should go defensive.

Chapter 7
Lumberyards

Lumberyards

Chapter 7

Introduction

A lumberyard is any facility or location where commercial lumber products are stored in large quantities for sale to the building construction trades or to the general public. There are two types of lumberyards today. The traditional lumberyard has one or more buildings on the premises and a significant amount of open space (Figure 7.1). The buildings house the sales area and often contain products that would be

Figure 7.1. Traditional lumber company.

damaged by exposure to the weather. The open space area contains lumber stacks of two-by-three-inch and larger lumber. The second type is the large retail store that merchandises miscellaneous building supplies (Figure 7.2). These are generally known as home improvement centers. They frequently store large amounts of lumber products that can create fires that can be extremely difficult to control. In addition, they frequently store large quantities of other products, such as flammable liquids, miscellaneous combustibles, and toxic materials, that can compound the problem when fires occur.

Figure 7.2. Home improvement centers.

Occupancy-Specific Cues

Construction and Content Factors

The construction and contents of lumberyards and home improvement centers may be quite different. They may vary widely not only in the construction-type of their buildings, but also in where products are stored or stacked.

Traditional Lumberyards. Structures in traditional lumberyards can be a factor in fire growth and spread if the fire starts in them or spreads to them. In most cases, however, lumber is stored outside, and structures on the premises are small, usually consisting of offices or sales areas and storage sheds.

Structures in traditional lumberyard facilities are usually of ordinary or wood-frame construction and will contribute to the fire load. Stability under fire conditions may be questionable in structures used to store lumber or wood products, and collapse under severe fire conditions may be rapid.

Metal buildings are used frequently in traditional lumberyard facilities for semiprotected storage. Although these structures may appear sturdy, they can collapse very quickly when subjected to the heat produced by burning lumber or wood products.

In addition to office structures in traditional lumberyard facilities, finished lumber products such as moldings, paneling, and plywood may be stored in open or closed structures. The value of the contents within structures must be considered during fire situations. In most cases, business records, finished lumber products, hardware items, and tools and machinery are much more valuable than raw lumber that may be burning.

Built-in fire protection usually is not provided for outside lumber storage. Alarm systems are found only occasionally in structures on the premises. For this reason, the reporting of fires in lumberyards is delayed frequently, resulting in large fires by the time the fire department reaches the scene.

Home Improvement Centers. In home improvement centers, the structures are usually ordinary (masonry walls with wooden roof assemblies) construction or noncombustible construction. Generally, they have very large floor areas. They are often the height of two-story buildings but have only one floor, and they have high-rack storage inside as well as outside. They may or may not have automatic sprinkler protection, depending on the local codes. The high-rack storage system produces a maze-like pattern. In smoke conditions, firefighters can become disoriented and lost in the structure.

Fires that start in the high-rack storage may spread very rapidly through the Class A combustibles. Smoke spread will be rapid due to the heavy fire loading and the wide-open area of the store. The racks, which may be constructed of wood or steel, are subject to collapse under fire conditions.

Common Problems

Access to both types of lumberyard facilities may be difficult, especially in terms of reaching the fire area. Approaching the fire area with apparatus may not be possible, resulting in long hose line stretches to reach the fire. Place fire apparatus carefully at traditional lumberyard facilities to account for possible fire spread or collapse of stacked lumber piles. Access points into and around the facility should be preplanned.

Water supply for fires in lumberyards can be a problem because hydrants may not be available in the area. Consideration must be given to the total fire flow required for a large fire in the facility. Even when area hydrants are available, the water supply may not be sufficient to support the heavy streams required for the fire potential.

Rapid fire spread is common in lumberyard and home improvement center fires if fires are not reported quickly and suppressed in their early stages. Incident commanders must remember that lumberyard and home improvement center fires can grow rapidly, requiring major changes in strategy and increased resource commitments.

The type and amount of material burning and how it is stored will affect fire growth and firefighters' ability to extinguish it. For example, tightly stacked large-dimension lumber will not burn as rapidly as loosely stacked smaller-sized lumber.

In traditional lumberyard fires, the amount of heat radiation and convection produced by the burning material must be considered. Large, free-burning fires will have a greater potential for exposing adjacent structures and will require additional hose lines for exposure protection. These fires also will make it more difficult to approach the fire area to apply water in the most effective manner.

Stream penetration into stacked lumber is important and is made more difficult if the lumber is tightly stacked or if heat radiation limits the approach to the fire area. These factors may necessitate straight-tip streams along with hose lines with spray nozzles for protection. In traditional lumberyards, elevated streams may be required to place the water on the burning material.

In traditional lumberyard fires free-burning lumber can generate large quantities of flying brands. Winds and thermal columns created by the heat from the fire can carry these brands great distances and create spot fires quite distant from the original fire area. Consider downwind patrols to minimize this potential for additional fires.

Fires in lumberyards and home improvement centers can become long-term incidents. The size of the fire, the amount of material burned, and mop-up operations can require extended on-scene commitments of resources. Personnel fatigue also can be a factor due to the amount of work required and the possibly hostile operating environment. For this reason, consider rehab needs along with the use of relief personnel.

Large, long-duration fires in lumberyards usually will require extensive resource commitments. The potential for this type of fire demands early consideration of resource needs. Additional alarms or mutual aid must be requested in a timely manner. Also consider specialized resources that may be needed, such as elevated stream apparatus or lighting equipment for nighttime operations.

Also evaluate interagency support early, and request it as needed. These incidents probably will require law enforcement personnel to deal with traffic problems in the area and utility service companies to terminate or control affected utilities. You also should consider establishing emergency medical services capability at the scene to treat any injured firefighters.

Overhaul of lumberyard and home improvement center fires can be extensive and difficult. The large amounts of tightly stacked, burned, and charred material present and the difficulty of reaching deep-seated fires for final extinguishment make overhaul time consuming and labor intensive. The instability of stacked lumber also may be a problem. Wetting agents such as "light water" can speed this process considerably. In some cases heavy equipment is required to move material for the overhaul. One solution to these problems is not to conduct an overhaul of the burned material but to post a fire watch until certain that the fire is completely out.

Cue-Based Predictions

- When the fire is of significant proportions, the defensive mode will be applicable from the arrival of the first apparatus. Protect exposures and establish a high-volume water supply.
- Intense radiant heat may prevent firefighters and apparatus from approaching the fire area to place effective streams on the burning area.
- Long supply lines often require significant numbers of pumping apparatus to deliver water to the scene.
- Lumberyard fires can turn into conflagrations quickly if large streams are not used or if sufficient resources are not called early. Flying brands are a problem downwind.
- Master streams and large hand lines are required to knock down the fire.

Incident Management Cues

Fires in lumberyard facilities often require multiple tactical operations. Fire attack, exposure protection, and ventilation are just a few of the activities that must be coordinated during the incident. Failure to coordinate tactical operations not only can result in very ineffective fire control but also can cause serious injury to fire fighting personnel.

Command Officers must ensure that tactical operations are performed in a timely manner and do not conflict with each other. Opposing hose lines and premature or delayed ventilation are examples of problems that can develop if tactical operations are not controlled and coordinated.

Strategy and Tactics

The basic strategy for controlling fires in lumberyard facilities will vary, to some extent, depending on whether the fire is in an outside lumber storage area or inside a structure.

Outside Storage Fires

Fires in outside lumber storage can grow quickly and spread rapidly depending on the type of materials involved and the method of storage. They can create spectacular blazes that generate enormous amounts of heat and spread flying brands over large areas.

The basic strategy for controlling an exterior lumber storage fire depends on the size of the fire, access to the fire, and proximity of any exposures. If the fire is not large, you can use small hose lines (1¾-inch) for a direct attack on the fire, with additional small lines placed to protect any exposed lumber storage or adjacent structures.

In the case of larger fires in outside lumber storage, you should deploy large hand lines (2½-inch) to provide the reach, fire flow, and penetration required to extinguish the fire. In some situations, the size of the fire may dictate the use of master streams from fixed appliances. If the fire involves highly stacked lumber storage, consider elevated streams. When employing large hand lines or master streams, consider the ability of the available water supply to support the operation.

If conditions permit, attack the fire from a direction that will not move it toward uninvolved material or toward adjacent structures. If structures are threatened by heat radiation or convection from the fire, place initial hose lines or appliances in positions to protect them. Flanking a hot, fast-moving fire will allow some fire control along with a measure of safety for firefighters.

As an Incident Commander you will need to establish Divisions/Sectors to control the exterior fire. An IC usually looks at the four sides of the facility and establishes a Division/Sector for each one (Figure 7.3).

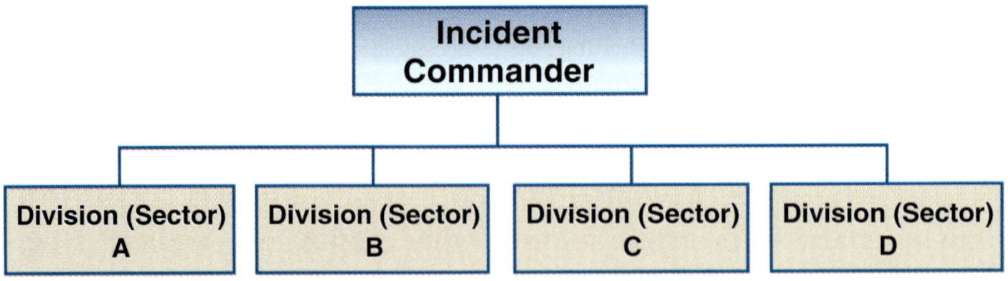

Figure 7.3. ICS organization at exterior lumberyard fire.

Safety must be a prime consideration when selecting strategies or tactics to combat fires in lumber storage. This is especially true in larger fires. The heat radiation produced and the possibility of collapse of stacked lumber are particular concerns.

Fires Inside Structures

Fires that occur inside structures of lumberyard facilities or home improvement centers present different problems than those that involve only outside storage.

If the fire occurs in the facility during normal business hours, life safety of employees and customers must be a primary consideration.

The amount of lumber or other combustibles inside a structure of a traditional lumberyard is not usually as great as that of outside lumber storage. However, the types of materials (moldings, paneling, plywood, and other products) often can produce intense fires that are extremely difficult to control. Additionally, the structure itself may contribute to the fire load. Home improvement centers, on the other hand, often have much more combustibles on the interior than will the typical lumberyard.

As with any interior fire, the best approach will usually be a quick, aggressive attack on the seat of the fire. This must be tempered with concern for conditions within the building, the amount of fire, and the risks involved in entering the building. From the ICS standpoint, this can be accomplished by establishing a Division/Sector 1.

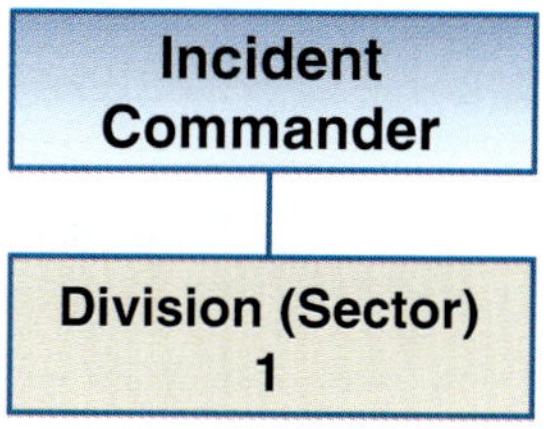

Consider vertical and horizontal ventilation of the structure as a means of providing a safe operating environment within the building and limiting the direction and amount of fire spread. This typically is accomplished by establishing a Vent Group.

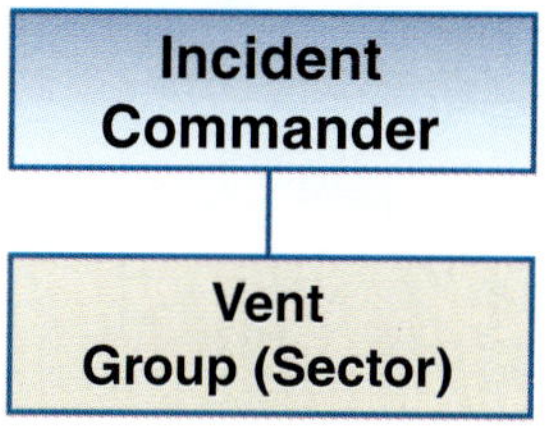

Also consider internal exposures, especially other combustibles, combustible liquids, or hazardous materials. Due to the large floor area of a home improvement center, you may need to divide the interior floor space between several Division/Sector Supervisors, and you may have to establish Exposure Division/Sector B and D.

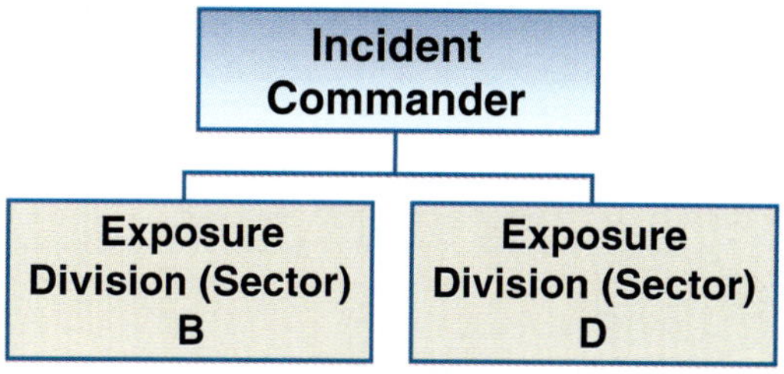

Salvage becomes a strategic consideration at fires involving structures in lumberyard facilities. Protection of building materials, hardware items, and tools from water and smoke damage can do much to limit the overall amount of fire loss. You can accomplish this by establishing a Salvage Group/Sector.

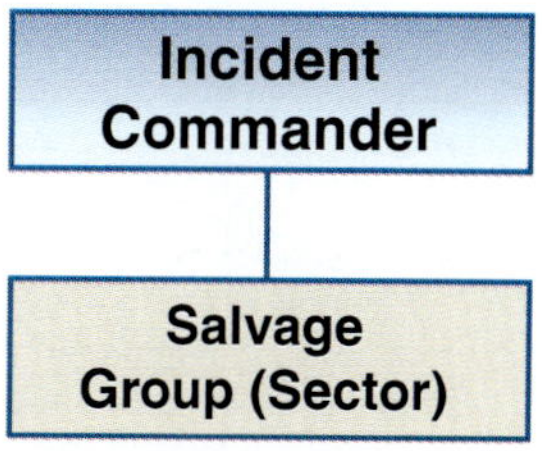

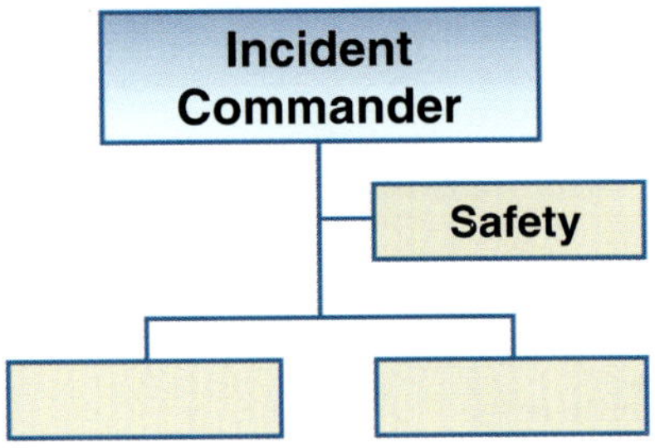

Safety considerations when fighting fires involving structures at lumberyard facilities are important. You must evaluate structural stability before firefighters enter the structure and monitor it constantly if the fire is not extinguished quickly. You will need to assign a Safety Officer in these situations.

Changes in Strategic Mode

Fires in lumberyards and home improvement centers, whether involving outside or inside storage, can develop rapidly and change quickly. Command officers should have alternate strategies in place to deal with these changing conditions. Initial placement of apparatus can be critical if the fire grows and necessitates a switch from an offensive attack with hand lines to a defensive operation with heavy streams. When a switch in operational mode is made, it is critical that all involved in the incident are made aware of the change and modify their operational status accordingly. The IC should obtain acknowledgment from officers directing tactical operations to prevent injuries or incompatible activities.

Activity 7.1
Lumberyards

Read each of the brief scenarios and then answer the questions that follow. For some questions, we have provided sample answers to get you started.

Scenario 1

A lumberyard is experiencing a major fire in a number of outside-storage lumber stacks. The fire is severely exposing other lumber stacks and one wood-frame storage building. What problems are created, and what tactical and Incident Command System (ICS) solutions must be applied?

Problem A: High radiant heat

Tactical Solution: Use master stream devices to knock-down the fire front.

ICS Solution: Establish Divisions/Sectors for proper supervision of operations.

Problem B:

Tactical Solution:

ICS Solution:

Problem C: ____

____

____

Tactical Solution: ____

____

____

ICS Solution: ____

____

____

Scenario 2

A lumberyard with a major involvement of one storage shed and several lumber stacks is seen on arrival. Describe your concerns for apparatus placement and personnel safety at this incident.

____

____

____

____

____

____

Situation 3

A lumberyard fire with a major involvement of nine stacks of outside lumber storage is seen on arrival. You estimate that you will need a continuous fire flow of 3,000 gpm from several master streams and hand lines to control the fire. The closest hydrants are 1,000 feet east and west of the lumberyard. The engines carry 1,000 feet of 4-inch hose. The static hydrant pressure is 55 psi. The maximum system flow is 5,000 gpm at 10 psi residual pressure. Given a standard alarm of 3 engines, 1 truck, and 1 chief, how many alarms would you need to handle this situation? Explain your answer.

____

____

____

____

Activity 7.1 Answers

Scenario 1

Problem A: High radiant heat

Tactical Solution: Use master stream devices to knock down the fire front.

ICS Solution: Establish Divisions/Sectors for proper supervision of operations.

Problem B: Exposures involved in fire due to radiant heat

Tactical Solution: Aggressively attack the structure fires to reduce property loss.

ICS Solution: Establish Divisions/Sectors to manage resources assigned to structure fires.

Problem C: Insufficient water supply.

Tactical Solution: Establish a water supply operation.

ICS Solution: Establish a Water Supply Group to ensure coordination and a sufficient water supply.

Situation 2

Apparatus can be severely damaged by radiant heat and must be positioned at a distance that will reduce the effect of heat on the unit. In addition, firefighters must be aware that positions that offer shielding from the radiant heat will be necessary when they are in close proximity to the fire.

Situation 3

Due to the long lays and reduced hydrant pressures, pumpers would have to be placed on each hydrant. In addition, to move the water required, additional apparatus would be required in a relay operation from each hydrant used. Pumpers would have to be on the scene to distribute the water effectively. This operation would take three to four alarms.

Chapter 8
Small Hotels and Boarding Houses

Small Hotels and Boarding Houses

Chapter 8

Introduction

A small hotel or boarding house, for the purposes of this chapter, is any building providing transient or short-term housing for fewer than 100 persons. Occupancies of this type are typically more than one but fewer than eight floors in height (Figure 8.1). Many occupancies referred to as boarding homes will fall into this category. This unit does not refer to large hotels that house more than 100 persons or are more than eight stories tall. Those occupancies typically are newer buildings that fit into the habitational high-rise category.

Occupancy-Specific Cues

The construction of small hotels and boarding houses will vary with their age, their design, and the building codes that were in effect when they were

Figure 8.1. Typical small hotel.

constructed. These types of buildings may be of ordinary (masonry, wood-joist), fire-resistive, or noncombustible construction, and in some isolated cases, may even be wood frame. It is not unusual to find buildings of this type that have been expanded or remodeled and are of mixed construction types. Current building and fire codes universally require fire-resistive or noncombustible construction for these types of occupancies.

Remember that construction type refers only to the building's basic structural elements and not to its furnishings or contents. In many cases, the original basic structure may have been modified during remodeling or renovation.

In terms of separation of space on individual floors, small hotels normally are well compartmented. This compartmentalization helps to contain a fire to the area of origin, but only if closed doors separate the area of fire origin from the rest of the building. Vertical compartmentalization in multistory hotels relies on protected shaft enclosures. This compartmentalization can fail if fire enters stair shafts, elevator shafts, or vertical voids that penetrate floors. It also is defeated when fire laps upward on the outside of the building from window to window above.

The type of construction and code requirements for built-in fire protection will have a dramatic impact on problems that are encountered in these occupancies. The built-in life safety protection will depend on the year of construction and the local building codes in effect at that time.

Fire Protection Systems

Alarm systems have to be considered the first line of defense in any habitational occupancy, and in small hotels and boarding houses they can range from nonexistent to state-of-the-art. Many times even though alarm systems are installed, they are not maintained and fail to function when fires do occur. Added to this, hotel staff personnel often are reluctant to report fires immediately and, thus, allow fires to grow until they are out of control.

Sprinkler systems are certainly one of the most effective built-in life safety features, yet they are found only in the newest structures of this type. In some cases, automatic sprinklers have been retrofitted under local or state ordinances that were passed following tragic fire losses. Often small hotels will have sprinklers installed only in selected areas, such as basements or hallways, but not in individual rooms. The National Fire Protection Association (NFPA) has developed a *Standard for Fire Department Operations at Properties Protected by Automatic Sprinkler Systems* (NFPA 13E).

Standpipe systems in small hotels usually are Class I systems (1½-inch) that are supplied from the domestic water supply system. Do not consider these systems reliable for primary fire department use. In most cases, they provide low working pressure and minimal water supply, and any equipment attached to them will not meet fire department standards. If Class I systems are used for initial attack, they should be backed up with regular fire department hose lines charged from a reliable supply source.

Class II standpipe systems (2½-inch) may be found in some newer small hotels. They are designed for fire department primary use and may be either wet systems or dry systems. Wet systems typically are supplied by a fire pump that provides much more water volume and pressure than a Class I system. Preplans should note the location of standpipe outlets in the building. When a Class II standpipe is present, it is the fire department's primary means of getting water to upper floors effectively.

Life Safety

Successful fire control and life safety operations in hotel occupancies require knowledge of the construction and physical layout of the building. Because of the potential for life loss, these types of occupancies should be preplanned periodically, and the preplan information should be available at the incident scene for use by Command Officers.

Because of the occupant load in small hotels, life safety is the major issue. Despite the term *small hotel,* the potential for life loss is great when compared to most other occupancies. The occupant load in a small hotel at any given time will vary according to several factors, such as the size of the structure, the time of year, and the time of day. Seven civilians die in every 1,000 dwelling fires in this country every year. By comparison, eighteen civilians die in every 1,000 small hotel and boarding house fires each year. Be thankful that there are many fewer small hotel and boarding house fires than dwelling fires.

The majority of small hotel and boarding house occupants are transients who lack knowledge of the building, exit routes, and any evacuation plan or instructions. In many cases, the life safety problem is compounded by the general behavior of the occupants of these structures, which frequently is not consistent with fire safety.

Access

Access to small hotels and boarding houses will vary depending on the design of the building and original code requirements (Figure 8.2). Normally a minimum of two means of ingress/egress are required. In some cases, the number of access points may be determined by exit requirements, based on occupant load. Access to all sides of the structure by fire apparatus may be hampered by parking lots filled with vehicles or narrow driveways and alleys. Because access points can be critical to fire control operations, they should be noted on preplans.

In addition to access to the building, access to the fire area may be delayed. This is especially true if the fire is on an upper floor of a multistory

Figure 8.2. Typical small hotel.

building. Smoke and heat in the hallways also may delay access to the area where the fire is burning. The horizontal distance that firefighters must travel, as well as the lengths of hose required to reach the fire area, can delay access in some cases.

The use of ground or aerial ladders to access the fire area may be difficult. Ground-level obstructions may prevent placing apparatus in positions from which ladders can be used. Utility poles and wires are common obstructions that often make ladder placement difficult or sometimes impossible. Many newer multistory structures may have inoperable windows that complicate entering the building from exterior ladders.

Multiple stair shafts are common in many multistory hotels. This requires a decision as to which stairways should be used for which purposes. You may need to select specific stair shafts for fire attack, ventilation, and evacuation. Also, you must remember that all stair shafts in the building may not provide access to the roof. This can cause serious problems when occupants attempt to evacuate upward in the building, and it also limits options for ventilating smoke-filled floors.

Open stair shafts, which can be found in many older occupancies, are particularly dangerous under fire conditions. They can allow immediate spread of fire and smoke on all floors in the building and complicate evacuation of occupants from the building.

Multistory small hotels usually are equipped with elevators that may be hydraulically or electrically operated. Both types are subject to system malfunctions that can cause them to respond to nonselected floors or become stuck in the hoist way. The safest approach to elevators is that fire personnel should not use them until it is determined that they are safe. It is essential that the fire department have control of the elevator lobby on the fire floor *before* using elevators for access in the building. A departmental procedure regarding use of elevators under fire conditions should be in place, and all personnel should follow it.

Water Supply

Water supply in areas where small hotels and boarding houses are located is usually from hydrants that will provide adequate flow for most fire fighting needs. If the fire involves several rooms or more than one floor, you must consider the ability of the hydrant supply to deliver the necessary fire flow. If a sprinkler system also must be supplied, assigning additional resources may be necessary to deliver the required fire flow.

Fire Development and Behavior

Fire development and behavior in small hotel and boarding house fires usually are no different than in other types of structures, as the fires typically are confined to a single room or living unit. Extinguishment can be relatively simple if doors to the fire area are closed and vertical lapping does not occur. However, if a door to the unit is open or fails, fire and smoke entering the hallway can create major problems. Any time that fire and smoke are allowed to enter hallways or stair shafts, serious life safety problems are created.

Vertical lapping presents the potential for fire on more than one floor (Figure 8.3). The rate of vertical extension will depend on several factors. The volume of fire exiting the area of origin certainly will be a factor, as will the type and thickness of glass and the window-frame material on upper floors. Single-strength glass set in aluminum frames can fail very quickly when exposed to flames from a fire below.

Figure 8.3. Multistory structures present the potential for vertical firespread through windows or other openings.

Additional Resources

You must consider resources in terms of fire condition and the amount of work that must be done. A small fire that does not require extensive search and rescue usually can be handled with the initial alarm assignment. These cases will require fire attack, limited evacuation, and possibly some simple ventilation. However, with a larger fire that can spread rapidly, coupled with a major search and rescue operation and complex ventilation requirements, you will need additional resources. If any doubt exists that initial alarm resources are adequate, request additional resources in a timely manner.

Medical Group/Medical Branch

A Medical Group/Sector or Medical Branch may have to be established at a multicasualty situation (Figure 8.4). If so, it may require the following units:

- Triage
- Treatment
- Transportation
- Medical supply
- Morgue

Some agencies establish a Medical Staging for patient transport vehicles under the Transportation Unit.

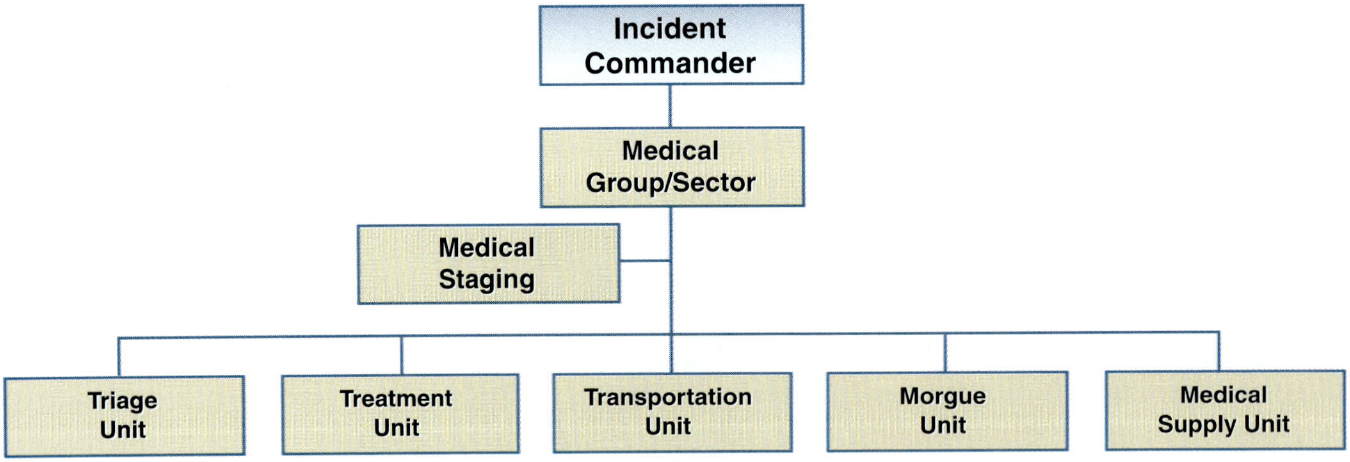

Figure 8.4. Medical group/sector organization.

Cue-Based Predictions

- A large number of people may become victims in any fire that produces large amounts of smoke.
- The ability to access the floors quickly may be hampered by open stairwells, people evacuating, limited number of stairways, or narrow stairways.

- Lack of a standpipe system may make fire attack more difficult.
- Many occupants are transients and are not familiar with egress.
- Fires often are contained to a single occupancy.
- A Medical Group/Sector or Medical Branch may be required to care for the injured.

Incident Management Cues

Working fires in small hotels and boarding houses frequently require performing several tactical operations simultaneously. Fire attack, search and rescue, and ventilation usually occur at the same time. To ensure personnel safety and effective fire control, these operations must be coordinated. The responsibility for coordinating tactical operations rests with the Command Officers who are directing or supervising them. Proper coordination requires not only good communication among the officers but also feedback from personnel who are directly involved in the specific operations.

Failure to coordinate tactical operations can result in serious problems. One result of poor coordination can be mixed attack modes where offensive and defensive operations occur at the same time in opposition to each other. Delayed or improper ventilation is another example of poor coordination that can have serious effects at incidents of this type. Uncoordinated tactical operations during small hotel and boarding house fires can jeopardize the lives of both building occupants and firefighters. Untenable conditions can be created within the building that may prevent evacuating occupants. Tactical operations may be impeded seriously, and smoke and fire can be spread throughout the entire building.

Strategy and Tactics

The basic strategy for fires in small hotels should be based on the principles of Incident Priorities:

1. Life Safety
2. Incident Stabilization
3. Property Conservation

Size-Up and Initial Assignments

An effective size-up that identifies the problems will be critical in determining the proper strategy. However, the problems may not all be evident from outside the building. Outside observations should be used as an indicator of problems inside the building until additional information can be provided. Conditions inside the building should be evaluated as soon as possible. In addition to life safety concerns, this information should include horizontal or vertical extension of the fire. Once the problems are identified, strategic and tactical solutions should be developed.

Identifying the problems is also the first step in determining what resources are required. In simple terms, match the problems with resources. To do this, you must understand the availability and capability of the resources that are on scene or have been requested. You must consider how long the requested resources will take to arrive at the scene and how much work they can do when they get there.

Base the assignment of initial alarm resources on which initial actions will do the most good. In most cases, an immediate and aggressive attack on the fire will ensure the safety of the largest number of building occupants. If an immediate attack on the fire cannot be started, evacuation of building occupants may be the best approach. Ideally, an immediate fire attack coupled with evacuation of those occupants in the most danger will work the best. Figure 8.5 illustrates one possible assignment for the first alarm of three engines, one truck, one medical unit, and one chief at a three-story small hotel. Engine 1 is attacking the fire on the second floor while Engine 3 does primary search. Engine 2, with a line, is searching the third floor. Truck 1 is setting-up positive pressure ventilation, and Ambulance 1 is assessing the need for additional medical resources.

You must consider how to use stairways for fire attack and occupant evacuation. Avoid doing both from the same stairway because it can impede the progress of firefighters and endanger civilians who are trying to exit the building. It is much better to designate one stairway for evacuation and another for fire attack. The stairway used for evacuation should be maintained free of smoke. You can do this by using a stair shaft with a roof opening to remove smoke from the stairway and by pressurizing the stair shaft from below.

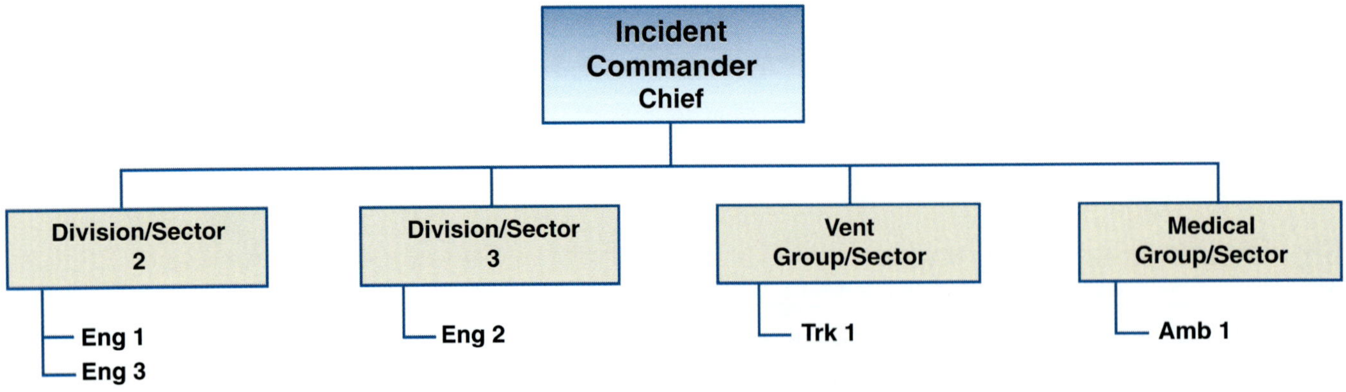

Figure 8.5. ICS for the first alarm at a three-story small hotel.

Fire Attack

The attack on the fire should be from a direction that does not push the fire into unburned areas of the floor. This can be difficult, depending on the location of the fire in relation to access points. However, positioning a separate hose line to limit fire extension often is possible, but you must ensure that this protective hose line is not used to attack the fire, which could create an opposing hose line situation.

For fires confined to one room or a small area, 1¾-inch hose lines normally are sufficient. A backup line also should be positioned as soon as possible. Search and rescue personnel should have a protection line when working in areas exposed to the fire. Heat buildup in hallways may require a large (2½-inch) hand line to absorb the heat and provide access to the fire area.

Search and Rescue

Search and rescue operations during fires in small hotels and boarding houses should consist of a primary and secondary search. The search should be started in the areas most affected by fire conditions. These normally will be the areas of the floor of origin that are in close proximity to the fire and the floors above to which heat and smoke have traveled. Occupants of the building should be directed to designated evacuation routes.

In some cases, depending on fire conditions, all occupants of the building may not have to be evacuated. If the fire is contained quickly and smoke removal is started early, many of the occupants will be in no danger and will be safe staying in their rooms. The decision not to evacuate must be based on fire conditions, good judgment, and the knowledge that the fire and smoke are being controlled.

Smoke and Heat

Smoke and heat from most fires that are confined to one room can be removed by simple horizontal ventilation. Placing smoke ejectors in the hallway and opening windows or sliding doors in the affected areas usually is sufficient. Positive-pressure ventilation also can be used effectively in some situations. One of the difficulties with this approach is that narrow hallways make it difficult to obtain the air seal needed if a single room is to be directly pressurized. An alternative is to pressurize the hallway from the stair shaft and open the door to the fire room or area.

Stair shafts in small hotels and boarding houses can be used to exhaust smoke from the building under certain conditions. The first requirement is that the stair shaft have a roof exit and that it be opened. Smoke ejectors can be used to channel smoke down hallways to the stair shaft. Take care that these operations do not jeopardize occupants who may be attempting to evacuate the building.

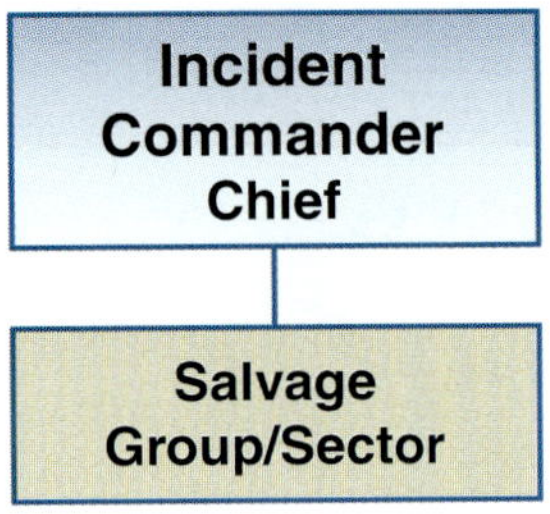

Salvage

Salvage operations should be started as soon as resources are available. Normally this will entail removing water from the fire floor and from the floor below and removing smoke from the entire building.

Personnel Safety

Safety issues associated with hotel fires must be considered. While back drafts are rare, they can happen, especially in tightly sealed rooms when fires have smoldered for a long time. For this reason care should be used when opening doors to the fire room or area.

The potential for flashover always exists with any structure fire, and hotel fires are no exception. Even if the fire is free-burning, flashover may not have occurred yet and may occur during the fire attack. Firefighters should not attempt entry into any area where a free-burning overhead fire is present until the area is cooled with hose streams.

Fatigue of fire personnel can be a safety factor when fighting hotel fires. The duration of the incident and the exertion of performing tasks at above-ground fires can quickly tire firefighters and lead to injuries. For this reason, periodic reliefs and use of rehab areas at extended incidents are recommended.

Activity 8.1
Small Hotel and Boarding House Questions

Directions

Read each of the brief scenarios and then answer the questions that follow. We have provided a sample answer for the first scenario to get you started.

Scenario 1

A 35' × 90', four-story boarding house of ordinary construction has a fire in the basement at 0100 hours. The structure has two open stairways from the first floor to the fourth. The interior basement entrance has a standard wooden door. The first floor is filling with smoke and heat. There are two basement windows on each side of the structure, as well as an exterior basement entrance at the rear. Approximately 30 residents are in the building.

What problems are created, and what tactical and Incident Command System (ICS) solutions must be applied?

Problem A: Fire in basement.

Tactical Solution: Ventilate the first floor to sustain operations and protect occupants. Attack fire by way of interior or exterior stairs. Ensure that upward extension through walls is controlled. If interior or exterior stairs fire attack is not possible, attempt to attack the fire from the basement windows or breach floor and use a distributor nozzle.

ICS Solution: Establish a Basement Division/Sector, a Division/Sector 1.

Problem B:

Tactical Solution:

ICS Solution: ____

____

____

Problem C: ____

____

____

Tactical Solution: ____

____

____

ICS Solution: ____

____

____

Scenario 2

A two-story bed-and-breakfast dwelling, 40' × 60', of frame construction, circa 1910, has a room and contents fire. Heavy smoke is seen on the first floor. The second floor has four bedrooms and a bath. A single open stairway leads to the second floor. Two people can be seen at a window on the second floor, Side B. Light smoke is issuing from that window.

What problems are created, and what tactical and ICS solutions must be applied?

Problem A: ____

____

____

Tactical Solution: ____

____

____

ICS Solution: ____

____

____

Problem B: ____

____

____

Tactical Solution: ____

____

____

ICS Solution: ____

____

____

Problem C: ____

____

____

Tactical Solution: ____

____

____

ICS Solution: ____

____

____

Scenario 3

A 60' × 80', three-story, small hotel has a fire in the elevator lobby area on the second floor. The hotel is fully sprinklered. Describe the actions that you would take to support the sprinkler system. What is your reference for this action?

____

____

____

____

____

____

Scenario 4

A one-story motel has 12 units in a row. The attic assembly is common to all the units. A severe fire is in the unit second from the end. The attic vents are showing heavy, dark smoke. What problems exist, and what tactical and ICS solutions must be applied?

Problem A: ______________________________

____

____

Tactical Solution: ______________________________

____

____

ICS Solution: ______________________________

____

____

Problem B: ______________________________

____

____

Tactical Solution: ______________________________

____

____

ICS Solution: ______________________________

____

____

Problem C: ______________________________

____

____

Tactical Solution: ______________________________

____

____

ICS Solution: ______________________________

____

____

Suggested Activity Answers
Activity 8.1

Scenario 1

Problem A: Fire in basement.

Tactical Solution: Ventilate the first floor to sustain operations and protect occupants. Attack fire by way of interior or exterior stairs. Ensure that upward extension through walls is controlled. If interior or exterior stairs fire attack is not possible, attempt to attack the fire from the basement windows or breach floor and use a distributor nozzle.

ICS Solution: Establish a basement Division/Sector, a Division/Sector 1.

Problem: People trapped above the fire.

Tactical Solution: Do a primary search of the entire structure and remove occupants from all floors.

ICS Solution: Establish a Rescue Group/Sector.

Problem C: Smoke in the structure.

Tactical Solution: Provide horizontal ventilation and vertical ventilation over the stair shafts.

ICS Solution: Establish a Vent Group/Sector.

Scenario 2

Problem A: Fire is most likely in a room on the first floor, given the smoke conditions. However, the fire could be in the basement, and that area must be checked immediately.

Tactical Solution: Make an attack on the fire. (For our answers, we will assume a first floor fire).

ICS Solution: Establish a Division/Sector 1 to manage operations of the fire attack and a Division/Sector 2 to manage extension control on the second floor.

Problem B: People trapped above the fire.

Tactical Solution: Remove the obviously trapped occupants and do a primary search of the rest of the building.

ICS Solution: Provide Divisions/Sectors 1 and 2 with sufficient resources to also do the search.

Problem C: Smoke in the structure, heavy on the first floor and probably lighter on the second floor.

Tactical Solution: Do horizontal ventilation, probably using positive-pressure ventilation.

ICS Solution: Establish a Vent Group/Sector

Scenario 3

Have one of the first alarm engine companies connect two 2½-inch hose lines (minimum size) to the automatic sprinkler system connection that supplies the building. Pressurize the hose lines at 150 psi engine pressure. This is to be done automatically after a company is assigned to supply the system. The applicable National Fire Protection Association reference is NFPA 13E.

Scenario 4

Problem A: Fire in a room.

Tactical Solution: Attack the fire in the room.

ICS Solution: Establish a Division/Sector 1

Problem B: Fire in the common attic space.

Tactical Solution: Confine the fire in the attic space by placing crews in the occupancies on each side and have them pull ceiling and direct hose streams into the attic area to extinguish the fire. Cut a vent hole over the fire occupancy.

ICS Solution: Establish an Exposure Division/Sector B, an exposure Division/Sector D, and a Vent Group/Sector.

Problem C: People possibly in other units.

Tactical Solution: Do a primary search.

ICS Solution: Establish a Rescue Group/Sector.

Chapter 9
Enclosed Malls

Enclosed Malls

Chapter 9

Introduction

An enclosed mall, for the purposes of this chapter, is any large retail complex contained under one roof, with the primary entrances to the various stores being from enclosed walkways (Figure 9.1). These occupancies are generally one or two stories in height and rarely exceed five stories. Malls higher than three stories usually are older buildings that have been renovated and turned into outlet malls or trendy retail shopping complexes. While the norm for this type of complex in most communities is from one to three stories, the philosophy that "bigger is better" does exist. The result of this philosophy is that in several communities across America "megamalls" are being constructed that may be a half-mile or more in length and contain mammoth anchor stores.

Figure 9.1 Typical one-story shopping mall.

Occupancy-Specific Cues

The construction of enclosed malls, like that of all occupancies, will vary with the building's age, original use, and design, and the building codes that were in effect when it was built. These complexes may be of noncombustible, mill or heavy timber, ordinary, or lightweight construction. However, it is most common to find mixed construction types in these complexes. Remember that the type of construction refers only to the basic structural elements of the original building and does not take into account its furnishings or contents.

Remodeled Structures

A popular trend for the past several years has been to remodel old office buildings, factories, pier complexes, and canneries into shopping malls. This renovation often modifies the original building design and may mislead fire personnel who are "reading" the apparent construction cues. As for all new structures, preplanning these types of buildings during construction is imperative, enabling firefighters to become thoroughly familiar with building specifics. If this is not done, the incident commander and fire crews may face a myriad of unexpected problems due to the remodeling. Building renovations become a critical issue in determining the fire fighting strategies and operational mode during an incident.

Fire-Protection Devices

Although modern building and fire codes vary from jurisdiction to jurisdiction, they generally require these complexes to have more exits, built-in fire protection systems, fire-resistive or noncombustible construction, and alarm systems. Compartmentalization and separations in modern malls offer much better fire protection, but ironically, the inherent design of enclosed malls provides for the storefronts to be open, negating fire protection to a large degree and promoting rapid horizontal spread of heat, fire, and smoke.

Unprotected openings are a serious problem in both modern and renovated malls. The open storefronts, wide walkways, open stairs and escalators, suspended ceilings, utility shafts, and other unprotected voids created during construction or renovation all contribute to the spread of fire and smoke. Since constructing a mall complex without creating void spaces is virtually impossible, every effort must be made to minimize their hazards. Whenever possible, void spaces should be limited in size by partitions extending from the floor to the roof. Heat, smoke, and fire detection equipment should be installed in the voids, ductwork, and shaftways to provide early detection. Automatic sprinkler systems also should be installed in these complexes. Remember that codes are subject to interpretation by contractors and installers; the local jurisdiction must carefully scrutinize the building plans and have a strong local fire code to avoid ending up with inadequate fire protection systems. Not even a sprinkler system will prevent catastrophic damage in a building if conditions allow the system to be overwhelmed.

Enclosed malls of modern construction will generally have Class II standpipe systems (2½-inch) with multiple outlets on each floor. These systems may be either "wet" or "dry" and are designed primarily for fire department use. Due to these shopping malls' size and complexity, the standpipe system becomes the primary means of supplying water for an effective interior attack. While these systems may initially be supplied with a fire pump in the building, a fire engine connected to a reliable water source should connect and pump into the system, providing necessary water and pressure. It therefore becomes important to understand what types of systems have been installed in the mall. Are they individual systems or cross-connected systems, and which connection serves which system? This information should be noted on the preplan, as should the location of standpipe outlets inside the complex.

Prefire planning is an essential element of successfully mitigating emergencies in enclosed malls. Included in this preplanning should be developing a knowledge of construction features and physical layout of the shopping center (Figure 9.2). A good working relationship with mall security and management is beneficial. Joint training will enhance the total operation should an emergency occur.

Figure 9.2 Preplanning should include learning the mall's construction features and physical layout.

Life Safety

Traditionally, mall fires have not resulted in a large loss of life. However, due to their occupancy load, maze-like design, and congestion, we can never rule out the possibility. On any given day, the difficulty of moving large numbers of people out of a mall and getting emergency equipment into the mall is tremendous. Depending on the weather, time of day, or the season, this problem can be compounded and become very labor intensive. While the numbers will certainly vary, you should anticipate occupants in the mall regardless of the time of response. Many stores conduct inventories or stock shelves after normal business hours; restaurants and movie theaters are open before or after normal business hours; and some malls have skating rinks, arcades, or walking tracks. Just because the stores are closed, you cannot rule out rescue.

Roof Assemblies

Modern shopping complexes often include lightweight truss construction as well as tile and metal roof coverings, all presenting unique problems for firefighters in terms of collapse and ventilation. Generally, total collapse is not a problem in mall fires. However, due to excessive fire loads, voids, and unprotected openings, mall fires can quickly overcome resources if a rapid, aggressive interior fire attack is not started on arrival. Lightweight roof construction, coupled with HVAC systems and other dead loads mounted on the roofs of modern malls, lends itself to partial collapse. Improper or illegal modifications, often involving structural members or affecting their structural integrity, can seriously affect fire spread and promote collapse.

Access

Access and egress must be considered when preplanning an enclosed shopping mall (Figure 9.3). Generally, designated exits are equipped with panic hardware and doors that swing outward, although this may not be the case in old converted buildings. Storefronts are usually covered with security mesh, while employee doors are metal clad and secured from the inside with a metal bar. These are considerations when planning forcible entry operations. Some storefronts are equipped will roll-up metal doors in

Figure 9.3 Typical shopping mall interior

Figure 9.4 Old commercial complex that has been renovated.

addition to the security mesh. These doors are designed to close automatically in case of fire. Depending on the circumstances, the doors may activate quickly, trapping less mobile customers inside the store and compounding the rescue problem.

Exiting customers, both on foot and in vehicles, will hamper apparatus access and laddering capabilities to the exterior. The mall design and location may also present access problems. The mall may be built on a pier, or one or more of its sides may face a body of water with no vehicle access. Visual enhancements, such as fountains, ponds, gardens, terraced entrances, and other design features, will hinder fire fighting efforts. Solutions to these problems must be worked out prior to the emergency (Figure 9.4).

Resource Considerations

Good pressure and adequate water supply are always important to fire suppression efforts, and especially so in an enclosed mall for several reasons. The size of most malls dictates long hose lays unless the fire is on the first floor near an entrance. Long hose lays are time consuming and require higher pressures to be effective. The design of malls tends to allow relatively small fires to quickly become large fires, requiring large hand lines, possibly even interior master streams. An Incident Commander must be familiar with the hydrant system and main sizes in the area as well as with any built-in fire protection systems on site.

An enclosed mall incident of any significance will require resources beyond a normal first-alarm assignment. Basically, when fighting a mall fire, you are dealing with a horizontal high-rise. Stretching long hose lines, moving necessary equipment, and checking the vast number of concealed spaces for extension all require a large number of personnel. Add to that the personnel demands of search and rescue, evacuation, responder rehab and relief, as well as overhaul, and the incident has rapidly exceeded the resource capabilities of many fire departments. Yet whether or not your department has the necessary resources, you face the same problems. Resource considerations of this magnitude must be preplanned. If, as an Incident Commander, you wait until the incident occurs to develop your resources, you will be too late.

Any operation of this magnitude will require a great deal of logistical support. Incidents in enclosed malls require much the same support as those in high-rise operations: breathing-air operations, portable lighting, movement of large amounts of equipment, and adequate relief personnel to name a few.

Incident Management Cues

Certain indicators, or cues, will always assist the well-informed incident commander in developing effective strategy and making valid predictions.

General Management Cues

Construction type and building design present key cues. One of the skills of an effective Incident Commander is the ability to "read" a building. For example, fire will probably spread more rapidly in renovated malls than in those of modern construction. Newer malls will probably contain some lightweight construction features as well as more efficient built-in fire protection. Based on the open stairwells, walkways, and storefronts, you can expect a rapid spread of heat, smoke, and fire, not only hampering fire suppression efforts but creating a serious salvage and overhaul problem.

The location of the fire is a major consideration and has a significant impact on the Incident Commander's initial strategic decisions. A fire in the basement or on lower floors has a much greater potential for vertical spread than does a fire on the top floor. Controlling a fire in the center of the complex, away from exits, is a more labor-intensive operation than controlling a fire on the perimeter of the complex adjacent to an exit.

Life hazard and rescue potential are always major considerations at enclosed mall incidents. However, time of day, weather conditions, holiday seasons, and available resources can all indicate the severity of the problem.

Complexity and Size

Due to the complexity and size of most enclosed malls and the potential for fire spread, the Incident Command Organization should be started immediately. If you determine that you have a working incident, you should consider the early establishment of branches for those functions that will probably utilize large numbers of resources, such as a Medical Branch, a

Suppression Branch, or a Search and Rescue Branch. Initially the branch may consist of only one engine company, but it is more efficient to feed resources into a branch, establishing divisions, groups, and sectors than it is to try to establish the branch later. In addition, with an event of this magnitude, divisions, groups, or sectors are likely to be established so rapidly that the span of control will quickly be exceeded. Using five as an ideal span of control, if you establish five divisions, groups, or sectors with a maximum of five companies in each, your span of control is reached with 25 resources. Utilizing branches, your span of control is increased to 125 companies, increasing your capability by a factor of five.

The physical size of the mall may require similar operations at multiple locations. For example, fire suppression activities may be required in several locations. Communications and accountability are simpler if the incident commander has to deal with only one Suppression Branch Director rather than with several division or group supervisors.

Coordinating tactical operations on an incident of this size is a major consideration. It is the responsibility of those officers who are directing or supervising the various operations. Effective coordination depends largely on good communications, including direction from command officers and progress reports from personnel performing specific tasks.

When tactical operations are not properly coordinated, serious problems can arise and firefighter safety may be jeopardized. Delayed or improper ventilation could cause crews to take unnecessary abuse or could allow the fire to spread before hose lines are in place. Opposing hose lines or mixed modes of operations may be employed, worsening the outcome and possibly causing injuries.

Life Safety

Firefighter safety should always be a concern for an Incident Commander. It becomes even more significant in mall fires because of their unique challenges. The operational area is very large, requiring long hose lays and extensive search commitment. This, coupled with extreme interior temperatures, will increase firefighter fatigue. Adequate resources and appropriate rotation of personnel will help alleviate this problem. Responder Rehab should be established early, and may have to be established in multiple locations due to the size of the complex. Rehab should have rehydration fluids available as well as medical personnel to check the firefighter's vital signs prior to releasing them back to staging.

Occupant safety also presents some unique issues. Granted, many people will "self-evacuate," get in their cars, and go home. However, you could easily have many occupants who require relocation or possibly shelter, depending on weather conditions. Again, due to the size of the complex, you may have to set up more than one Medical Group providing triage, treatment, and transportation. All of these groups would function under the Medical Branch Director. Depending on the situation, establishing a separate staging area for medical resources may be feasible.

Strategy and Tactics

The strategies and tactics developed for fires in enclosed malls, like any other fire, must be based on incident priorities:

1. Life Safety
2. Incident Stabilization
3. Property Conservation

Thorough preplanning and an effective, on-going size-up will assist in correctly analyzing the problems and formulating appropriate strategies. While outside observations are critical, all of the problems may not be obvious from the exterior, requiring an investigation to gain necessary additional information.

Problem Identification

The investigation to evaluate the interior should be completed as soon as possible. At a minimum, it should include the extent and location of the fire or other problems and life safety concerns. You cannot develop strategic and tactical solutions until you have identified the problems.

Problem identification also will assist the Incident Commander in determining the type and quantity of resources required. In order to match the resource to the problem, you must understand the availability and capability of obtainable resources. To determine this, you must consider both response time and the time it takes to place the resource in operation. This becomes especially important when dealing with mutual-aid resources that must travel great distances. A resource's crew may need to be fed and rested before going into combat, which delays the availability of that resource.

Fire Attack Operations

The first alarm companies should be assigned to those actions that will have the greatest impact on overall mitigation. If conditions allow, an aggressive attack on the fire will provide the best opportunity for confinement and enhance the safety of the greatest number of occupants. During hours of operation, evacuation must be started at once. Ideally, if resources permit, ventilation and fire attack should begin simultaneously.

If at all possible, fire attack operations should be coordinated with rescue efforts to avoid using the same stairways or exits, which can seriously impede both evacuation of occupants and fire fighting operations. For example, if fire suppression efforts are initiated from the south end of the mall, evacuation should be through the north exits if possible.

Due to the open area, care must be taken to avoid pushing the fire into uninvolved areas. It is critical to place hose lines in uninvolved areas on either side of the fire to prevent extension into voids. If the mode of operation is offensive and/or you are faced with resource limitations, 1¾-inch hose lines usually are sufficient for initial attack operations. However, because of the large open areas and the potential for rapid extension, you should back up these lines with large hand lines as soon as possible.

Ventilation is especially difficult in enclosed malls because of the lack of compartmentalization. The proper use of the HVAC system will greatly assist ventilation efforts. If the building engineer or other qualified personnel are not available, or if you have any doubt that the system will be effective, *do not use it.* Some malls have skylights that open automatically or other built-in ventilation devices.

Remember, firefighter fatigue and rehab are major concerns during enclosed mall operations. All facets of this type of operation are very labor intensive; adequate resources must be available and sufficient care for the number of personnel must be on hand.

Activity 9.1
Enclosed Malls

Directions

Read each of the brief scenarios and then answer the questions that follow. We have provided a sample answer for the first scenario to get you started.

Scenario 1

A one-story, 600 × 200 foot *enclosed mall* of ordinary construction has a fire in one of the stores. The store is in the middle of the mall and is about 30 percent involved. The fire and smoke extend into the center corridor area of the mall. The time of the alarm is 1500 hours on Thursday.

What problems are created and what tactical and Incident Command System (ICS) solutions must be applied?

Problem A: A retail occupancy, in the middle of the mall complex, with 30 percent involvement.

Tactical Solution: Attack the fire through the entrance nearest the involved occupancy. Ensure that both horizontal and vertical extension of fire is controlled. Ventilate the main public areas to support fire fighting operations and facilitate evacuation of occupants.

ICS Solution: Establish a Fire Attack Group/Sector

Problem B: ____________________

Tactical Solution: ____________________

ICS Solution: ____

____

____

Problem C: ____

____

____

Tactical Solution: ____

____

____

ICS Solution: ____

____

____

Problem D: ____

____

____

Tactical Solution: ____

____

____

ICS Solution: ____

____

____

Scenario 2

A two-story, 800 × 300 foot *enclosed mall* of ordinary construction with a lightweight steel truss roof structure has a fire in a toy store located on the first floor. This occupancy is about a third of the way down the mall. It is 0100 hours on a Sunday, and the store is approximately 25 percent involved. What problems are created and what tactical and Incident Command System solutions must be applied?

Problem A: ______________________________

____

____

Tactical Solution: ______________________________

____

____

ICS Solution: ______________________________

____

____

Problem B: ______________________________

____

____

Tactical Solution: ______________________________

____

____

ICS Solution: ______________________________

____

____

Problem C: ______________________________

____

____

Tactical Solution: ______________________________

____

____

ICS Solution: ______________________________

____

ICS Organization

Draw an ICS Organization chart that might be used for the incident described in Scenario 2.

Activity 9.1
Suggested Answers

Scenario 1

Problem A: A retail occupancy, in the middle of the mall complex, with 30 percent involvement.

Tactical Solution: Attack the fire through the entrance nearest the involved occupancy. Ensure that both horizontal and vertical extension of fire is controlled. Ventilate the main public areas to support fire fighting operations and facilitate evacuation of occupants.

ICS Solution: Establish a Fire Attack Group/Sector

Problem B: The large number of people likely to be shopping in that occupancy at that time of day, including children, elderly, and disabled customers.

Tactical Solution: Conduct a thorough primary search of the entire mall complex and remove occupants to a safe area.

ICS Solution: Establish a Rescue Group/Sector. This may later become a Rescue Branch if the rescue effort escalates. Remember this will be a very labor intensive operation.

Problem C: The amount of smoke in the occupancy. The design of enclosed malls allows smoke (as well as heat and fire) to spread rapidly throughout.

Tactical Solution: Provide horizontal ventilation, especially in the public areas. Provide vertical ventilation over the fire area if possible.

ICS Solution: Establish a Vent Group/Sector

Problem D: Potential for injuries, numbers of very young and very old that need assistance, and separated families.

Tactical Solution: Provide on-scene medical care and transportation to medical facilities if necessary. Provide some type of accountability/tracking for civilians; if fire/EMS personnel are not available to do this, perhaps volunteer workers could be utilized.

ICS Solution: Establish a Medical Group/Sector

Scenario 2

Problem A: Store on the first floor with 25 percent involvement.

Tactical Solution: Attack the fire through the entrance nearest the involved occupancy. Ensure that both horizontal and vertical extension of fire is controlled. Smoke will spread rapidly, so the structure must be ventilated to facilitate fire fighting and search efforts.

ICS Solution: Establish a Division/Sector 1 and Division/Sector 2.

Problem B: Quantity and density of smoke throughout.

Tactical Solution: Provide horizontal ventilation beginning on the first floor and then progressing to the second floor. Provide vertical ventilation using building design features (skylights, ventilators, exhaust systems, etc.). It is critical that ventilation is coordinated and that the ventilation method employed does not spread the fire.

ICS Solution: Establish a Vent Group/Sector.

Problem C: While the general public (shoppers) does not present a rescue problem due to the time of day, this does not mean no rescue problem exists. A number of employees (security, housekeeping, stock clerks, food service people, etc.) very likely may be in the complex.

Tactical Solution: Conduct a primary search on the entire complex, including maintenance corridors, stock rooms, break rooms, and other nonpublic areas.

ICS Solution: Establish a Rescue Group/Sector

ICS Organization:

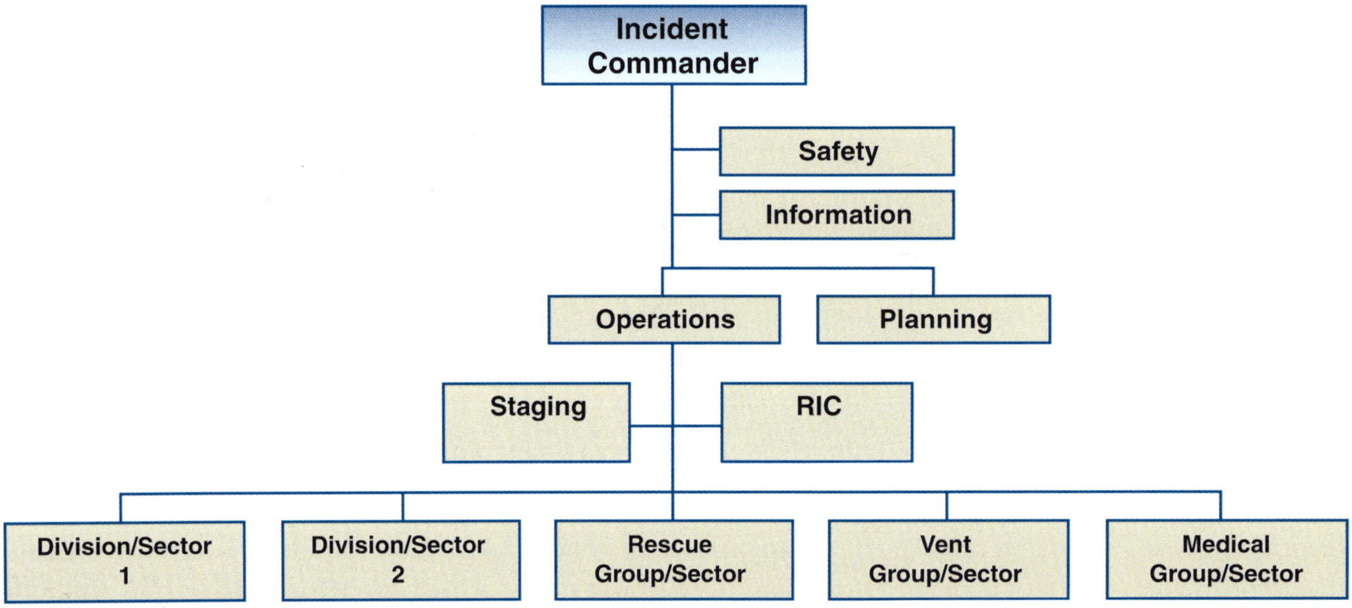

Chapter 10
Safety

Safety

Chapter 10

Introduction

This chapter focuses on the Command Officer's responsibilities in the command process and the impact of the command process on incident scene safety. The Command Officer's ability to manage safety concerns must be directly proportional to the probability of injury or death for emergency responders or the civilian population.

The IC in Safety Management

The IC is responsible for monitoring and assessing hazardous and unsafe situations and developing measures for assuring personnel safety. The IC must be aware of active and developing situations and evaluate the risk-benefit value of the various operations. Safety management requires a safety orientation in all we do. The IC must be a role model for safe operations and safety awareness for all personnel. Safety is an attitude, and it is essential that the IC have this attitude.

Safety SOGs

All personnel must operate safely at an incident. Every responder must be familiar with all departmental safety orders and standard operating guidelines (SOGs). Personnel must be aware of what is occurring in their area of operation and constantly be on guard to prevent injuries.

Incident Attire Requirements

Every department must have a set of written regulations and SOGs that establish baseline requirements for appropriate incident attire (Figure 10.1):

Figure 10.1. What are the safety concerns?

- Turnout coats, pants, and boots
- Approved helmets, gloves, and hoods
- Positive-pressure self-contained breathing apparatus (SCBA)

Operational Requirements

Regulations and SOGs must address the following operational requirements:

- A team approach to tasks, especially on interior and roof operations
- Sufficient supervisory personnel; one supervisor for every crew operating in a geographic location or functional responsibility
- Sufficient portable radios; one for each supervisor
- Sufficient hose lines and backup personnel; enough fire flow to control the fire and backup lines for safety
- Sufficient reserve personnel in Staging or at the Command Post (CP); one or more crews in Staging, should something go wrong
- A Rapid Intervention Crew (RIC)
- Safety Officer position staffed on working incidents
- A responder rehab area with fluids and an EMS unit

Enforcement of Safety Regulations

Personnel who violate safety procedures must be disciplined even if they are injured while violating a safety rule. Appropriate, firm punishment should be administered immediately after they return to work. The failure to enforce safety regulations indicates that management does not care about its personnel. Failure to enforce safety rules makes the rules ineffective.

Risk-Benefit Perspective

The Incident Commander must continually evaluate the risk being assumed by firefighting personnel versus the benefit of their achievements. A certain amount of risk comes with the job, but it is prudent to compare the degree of risk to the potential for success.

Risk Policy

Just as departments should have a clearly stated safety policy, so should they have a clearly understood risk policy. Command officers should thoroughly understand risk/benefit and what constitutes unacceptible risk.

A risk policy example:

We will take great risk to save your life.

We will take minimal risk to save your property.

We will take no risk for lives or property that are already lost.

This example is both a safety philosophy and a policy. It is easy to understand, recall, and apply. With such a policy, anyone can raise the issue of poor risk-benefit decisions at the incident scene when orders seem to be in opposition to the policy.

Risk-Benefit Cues

The well-informed Incident Commander will consider every available cue when making risk-benefit decisions and never allow complacency to enter into the decision-making process (Figure 10.2).

- Resource Cues that Affect Decision Making:
  - — Is sufficient water available at the scene to fight the fire and to provide backup lines?
  - — Are sufficient personnel available for the number of jobs? When personnel are limited and the number of tasks is large, command officers must set priorities based on value judgments of which actions provide the best benefit.
  - — Is the available apparatus or equipment appropriate for the job at hand?

Figure 10.2. Structural integrity must remain a concern.

- Apparatus Placement Decision-Making Cues:
  - — Defensive fire operations; volume of fire; structural deterioration; collapse probability
  - — Offensive fire operations
  - — Hazardous materials incident; type of chemical, danger to personnel
  - — Properly trained and equipped personnel
- Rescue Decision-Making Cues:
  - — Firefighters working above fire; only one egress; fire conditions below
  - — Open stairs; fire, smoke, or toxic gases threatening the egress
  - — Victims in area of flashover will perish in one minute or less
  - — Smoke conditions; inability to maneuver quickly; possibility of becoming lost
  - — Type of hazardous material involved; its chemical classification and consequent hazard to personnel
- Exposure Protection Decision-Making Cues (Figure 10.3):
  - — Magnitude of fire in building of origin; proximity of the exposure to the fire building; narrow streets or alleys
  - — Collapse condition of building of origin; deterioration of the structural members; number of floors involved
  - — Degree of involvement of exposed building

Figure 10.3. Look for cues that will guide your exposure protection decisions.

Figure 10.4. What's burning? What's the threat?

— A separate fire, not just an exposure problem

— Sufficient resources available

- Confinement/Extinguishment Decision-Making Cues (Figure 10.4):

  — Flashover imminent; rollover occurring; reports from the interior; smoke and heat conditions

  — Back draft imminent; dark, thick smoke; oxygen-starved fire; building "breathing"

  — Backup line in place; no backup line—interior crews at higher risk should conditions change

  — Smoke conditions; interior personnel cannot see well or move rapidly; personnel may get disoriented and lost

- Overhaul Decision-Making Cues:
  - — Condition of the structure or facility; weakened or damaged floor, wall, or roof assemblies; chimneys
  - — Amount of products of combustion remaining; any smoke or invisible toxic gases, wear masks
  - — Firefighter fatigue; lowered awareness, slowed reactions
- Ventilation Decision-Making Cues:
  - — Pulling or pushing fire on interior operating forces; where are the combustion products being directed; who or what is in their path?
  - — Firefighters working above the fire; open stairwells; smoke conditions; secondary egress
  - — Sufficient hose lines and water to control a vented fire; flare up; flashover; increased British thermal unit (Btu) production
- Salvage Decision-Making Cues (Figure 10.5):
  - — Products of combustion present; wear masks in smoke or invisible toxic gas atmospheres
  - — Fire location and direction of spread; would salvage crews be in jeopardy?
  - — Collapse potential of the structure; fire intensity; structural deterioration

Figure 10.5. A common attic can "hide" large volumes of fire.

Risk-Benefit Procedures

All personnel must be self-disciplined and make practical risk-benefit judgments regarding life safety. Self-discipline is critical to firefighter safety. Poor risk-benefit operational decisions must be brought to the attention of the immediate supervisor. Nearly everyone is a victim of tunnel vision from time to time. Personnel should question orders that appear to have too great a risk. This questioning will allow the supervisor to rethink the risk-benefit, thus possibly overcoming tunnel vision.

Anticipating Safety Factors in an Escalating Incident

Murphy's Law often applies on the incident scene ("If something can go wrong, it will"). Command Officers must keep their eyes on the big picture. They must be constantly aware of the variables at the entire incident scene. The only way to anticipate problems and solutions is to gather information, looking for what could go wrong.

Cues that indicate additional spread or structural deterioration include:

- Flames or dark smoke pushing from previously clear areas or vents
- Increased flame production
- Cracks in walls; horizontal crack may indicate floor sagging
- Inordinate time to control fire with what appear to be sufficient streams
- Subcommanders continually requesting additional crews and lines

Building Collapse Indicators

There are many reasons for building collapse during fire fighting (Figure 10.6). Critiques of various collapses have shown that only after a collapse has occurred was it realized that seemingly unrelated occurrences (or indicators) contributed to the collapse. The probability of only one indicator or defect causing the collapse was remote, but when combined

Figure 10.6. Understand your building.

with other defects, they caused a devastating problem. The prompt reporting of any collapse indicators to the IC will allow him or her to form a clearer picture of the building's stability.

The indicators of building collapse include:

- Continued or heavy fire
- Smoke showing through walls (may precede a back draft)
- Fire showing through a wall
- Back draft; heavy smoke and high heat coupled with inadequate ventilation
- Cracking noises coming from a building
- Sagging floors
- Bulging walls
- Leaning walls
- Failure of part of a wall
- Interior collapse
- Visible spalling of a brick wall
- Spalling of concrete and exposure of steel rods
- Wall breaking down under a hose stream
- Water between bricks
- Two or more floors of a building fully involved
- Excessive water in a building
- Water not coming out of a building as fast as it is going in
- Absorbing bales in a building
- New wall cracks
- Old wall cracks enlarging
- Severely dry-rotted or termite-eaten supporting wood beams exposed to heavy fire
- Expansion of structural steel beams
- Unprotected steel columns and beams exposed to heavy fire
- Large machinery in a building
- Plaster sliding off of walls in large sheets
- Excessive snow or water on a roof
- Numerous previous fires in a building
- Wall spreaders (The wall spreader is used in a building of ordinary [masonry, wood-joist] construction to pull a wall back to its original alignment. It is normally indicated by a "star" or rectangular plate on the exterior wall [Figure 10.7].)

Figure 10.7. Wall separator stars are common in older construction.

This list is by no means complete. If you are in charge and the above signs are noticed, you should take immediate action. Waiting for a variety of warnings and not acting immediately could result in your final warning or indicator being a collapsing building. The proper action will depend on whether any civilian life is in jeopardy. If all civilians are safe, then firefighter safety is your number-one consideration.

Safety Officer concept

While all firefighters should be their own "safety officer," there must be someone on the incident scene who has the primary responsibility for safety.

Structure Incidents

When to Staff. When should you staff the Safety Officer position? Some departments staff the Safety Officer position on every working fire. Others staff it when the second alarm is called. Departments that have limited personnel staff it on a sporadic basis. Ideally, a Safety Officer should respond to and operate at every working fire.

Safety Officer Qualifications. The Safety Officer must be knowledgeable of fire behavior, including fire's effect on structures, facilities, and the environment.

The Safety Officer must understand building construction, including how and when fire or heat affects the bearing members. As an example, unprotected steel members weaken at about 1,100° F *(593° C)*, and wood-truss assemblies can fail in as little as five minutes. Safety Officers also must understand how fire spreads through a given construction type. Balloon-frame construction is an example. Safety Officers must understand the collapse potential of the various construction types. Noncombustible construction is the most prone to collapse.

The Safety Officer also must know departmental safety and operating regulations and SOGs, as well as local, state, and federal laws governing safe operations for incidents to which the fire department responds.

Authority of the Safety Officer. The Safety Officer works directly for the IC and has full authority to correct actions or to stop poor risk-benefit operations. When personnel are determined to be in imminent danger, the Safety Officer has the authority to stop the operation. This is true regardless of the relative ranks of the Safety Officer and any other person involved in the action or operation. In this context, *imminent* means right now, with no time to go through channels. The Safety Officer perceives that within the time that it would take to notify the IC to stop the action or operation, personnel will be injured or killed. In other situations, the Safety Officer will notify the IC and have the IC take the corrective action.

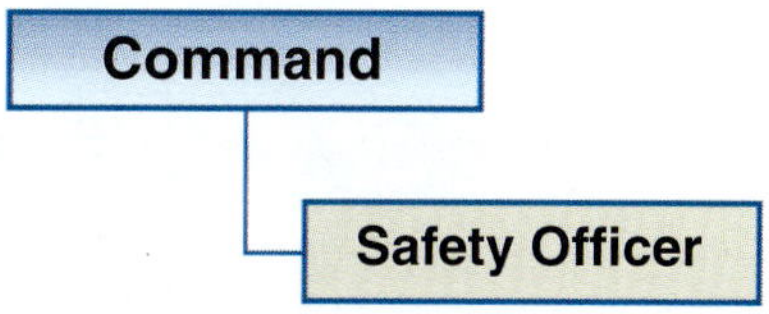

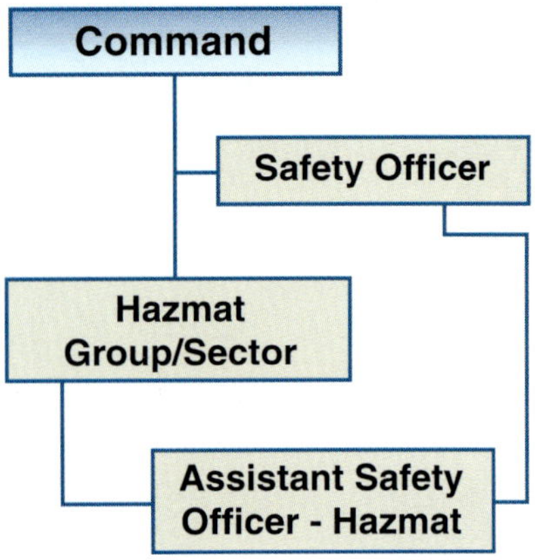

Hazardous Materials Incidents

The Hazmat Safety Officer oversees all hazmat-related operations. CFR 1910.120–OSHA Regulations states:

> The individual in charge of the ICS shall designate a Safety Officer, who is knowledgeable in fire fighting or rescue and hazardous substance handling procedures, with specific responsibility to identify and evaluate hazards and to provide direction with respect to the safety of operations of the emergency at hand.
>
> When activities are judged by the Safety Officer to be unsafe and/or to involve an imminent danger condition, the Safety Officer shall have the authority to alter, suspend, or terminate those activities. The Safety Officer shall immediately inform the individual in charge of the ICS of any actions taken to correct hazards at an emergency scene.

U.S. Fire Administration Firefighter Death and Injury Statistics

Using firefighter injuries per 1,000 fires, where are firefighters injured? The chart in Figure 10.8 illustrates illustrates the locations where firefighter injuries occur.

Most Common Activities. As can be seen in Figure 10.9, the three activities that contribute the most to firefighter injuries are:

- Handling charged hose lines
- Extinguishment, other (this means that the injury was sustained during some activity other than the ones listed)
- Overhaul

Leading Causal Factors. As responsible managers we must evaluate data developed by NFPA, OSHA, and others in addition to analyzing our own performance and safety history. By doing this we can take a proactive approach, developing improved policies and procedures or replacing equipment that will lead to a reduction in firefighter death and injury (Figure 10.10).

Responding to the Statistics. What actions should the fire service take based on this statistical information on firefighter injuries?

- Increase, or begin, prefire planning for the occupancies that have high firefighter injury rates. Attempt to identify hazards or hazardous areas.
- Increase the risk-benefit analysis for the designated occupancies.

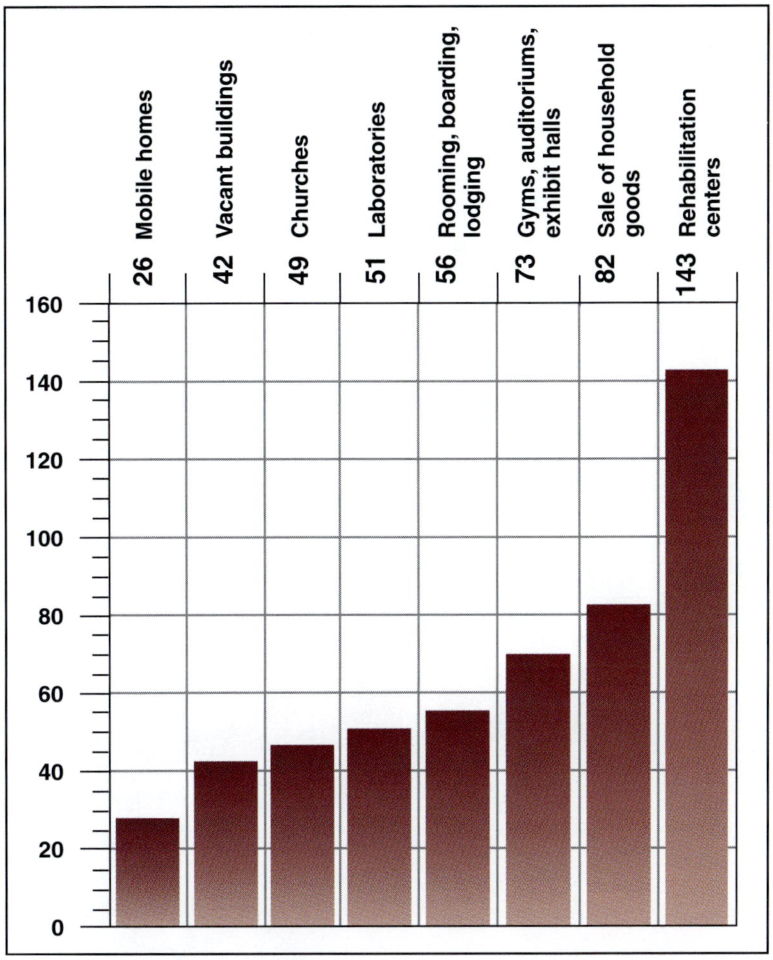

Figure 10.8. Locations of firefighter injuries per 1,000 fires.

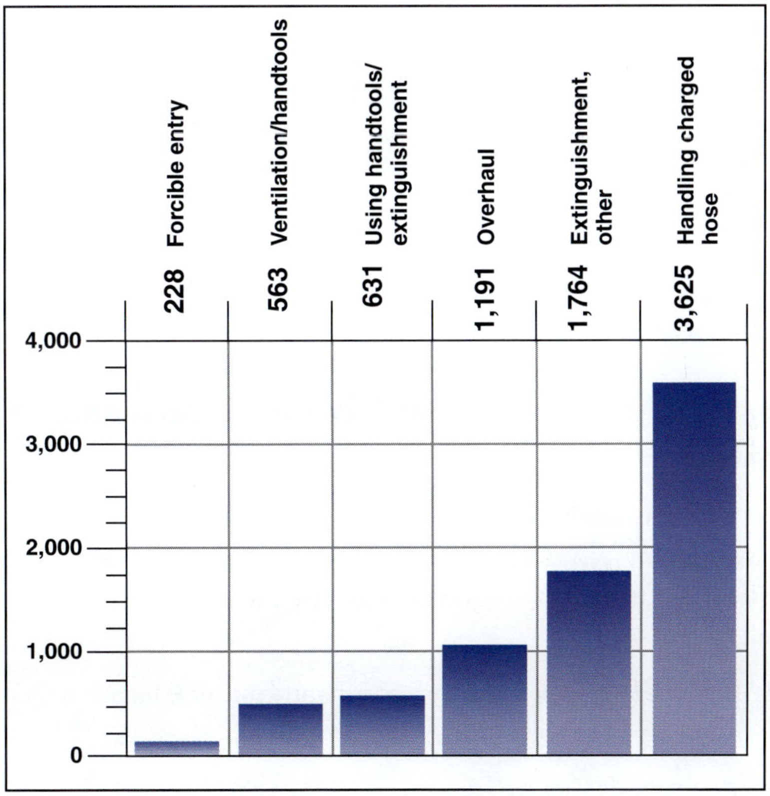

Figure 10.9. Causes of fifefighter injuries by activity.

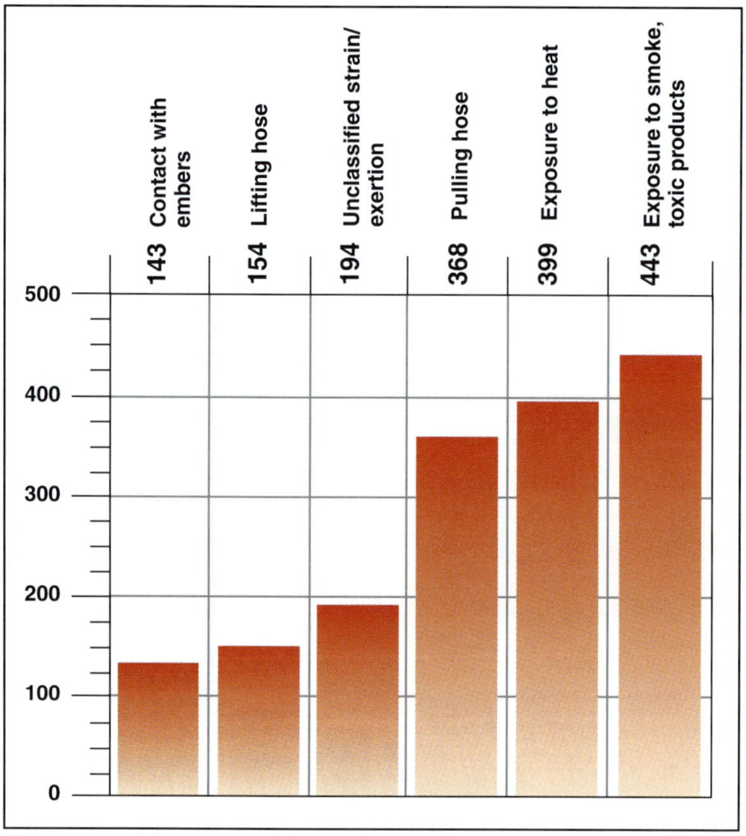

Figure 10.10. Leading causal factors in firefighter injuries.

- Change from offensive to defensive operations earlier.
- Ensure that, should operations turn defensive, personnel are far enough away from the structure.
- Increase awareness of all personnel, especially Command Officers, of the statistical information and how that information relates to incident scene safety.
- Ensure that firefighters are properly attired in SCBA to prevent smoke and toxic product exposure.
- Ensure that firefighters are in full protective clothing to prevent burn injuries.

U.S. Fire Administration Civilian Death and Injury Statistics

Using civilian deaths per 1,000 fires, where do civilian deaths occur (Figure 10.11)? What actions should the fire service take based on the statistical information for civilian death rates at mobile homes and boarding houses?

- Increase fire prevention inspections and public education focused on residents of mobile homes and boarding houses.

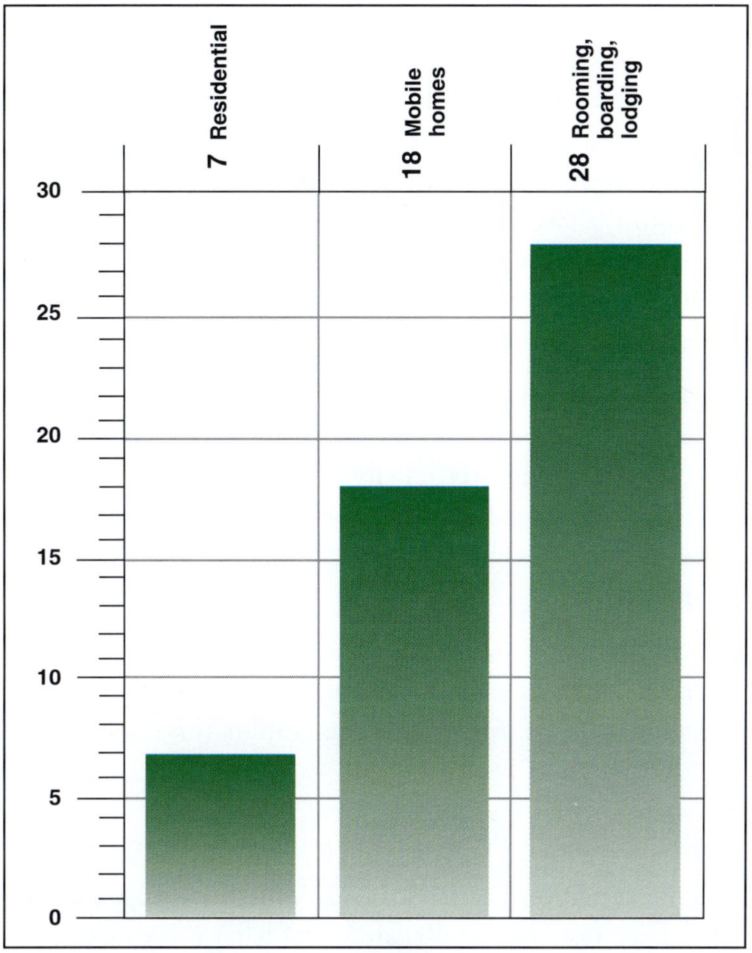

Figure 10.11. Locations of civilian fire deaths (per 1,000 fires).

- Increase the response of personnel, apparatus, and equipment to mobile home and boarding house fires.
- Decrease response time to mobile home and boarding house fires.

Safety Issues

Two-In/Two-Out Rule

The two-in/two-out rule is an often confusing and controversial provision of the OSHA Respiratory Protection Final Rule (29 CFR 1910 & 1926, January 8, 1998). This recently revised standard requires that any interior structural fire fighting procedures provide for at least two firefighters inside the structure. These firefighters must be in direct visual or voice communications with each other as well as in direct visual or voice contact with firefighters outside the structure.

It is unfortunate that not all firefighters are covered by the OSHA respiratory protection standard. However, it is prudent to consider the two in/two out requirements to be the minimum acceptable standard for safe fire ground operations for *all* firefighters when self-contained breathing apparatus is used. This standard is mandatory in those states that fall under either state or federal OSHA regulations.

The outside team must be in place and properly equipped *before* the attack team enters the structure. Considering the staffing levels of many departments, following this standard will certainly affect personnel assignments, capabilities, and possibly even your mode of operation. But keep in mind that *no building* is worth a firefighter's life. The only permitted deviation from the two-in/two-out rule is when first-arriving firefighters find a *known rescue situation* where immediate action may prevent the loss of life. This deviation must remain the exception and not be allowed to become the norm.

The outside team must be prepared for immediate assignment. However, the outside firefighters may perform other tasks, as long as those tasks do not interfere with their responsibility to account for the inside team. Any function performed by the outside crew must be able to be abandoned, without placing another employee in jeopardy, should a rescue be necessary.

For additional information on this subject, contact the International Association of Fire Fighters or the International Association of Fire Chiefs. It is important that you understand and employ this concept.

Rapid Intervention Crew

The NFPA calls it a Rapid Intervention Crew (RIC). In some departments it's a Rapid Intervention Team (RIT). Or it might be a back-up entry team or an emergency rescue crew. Regardless of what it is called, it's a critical function, and it is vital to recognize its importance to firefighter safety and survival.

Some departments have sufficient resources to build a dedicated rapid intervention crew into the dispatch or to assign a specific company (or companies) on arrival (e.g., third-due engine). However, most of America's fire service is comprised of small paid, combination, or all-volunteer departments that operate with very limited staffing. While resource limitations obviously will affect how the RIC is staffed or who is assigned to it, a responsible incident commander will not allow these limitations to preclude its establishment. He or she must be guided solely by the safety needs of the incident.

Why Have an RIC? It is well recognized that firefighters are exposed to the greatest risk of injury or death while operating at the scene of an emergency. As responsible managers, reducing that risk should be our primary concern.

Of course, it is unrealistic to assume that firefighters can avoid, control, or eliminate all risk from their day-to-day operations. Danger is an integral part of their work environment, and firefighters will always view an incident with some pessimism, well aware that things can go wrong. Recognizing this, we must make provisions to assist emergency personnel who find themselves in jeopardy. One of the most effective mechanisms for doing this is developing and training a rapid intervention crew.

One of the IC's most important responsibilities is conducting an early risk-benefit analysis of the incident in order to identify risk characteristics of the situation and the dangers related to specific activities. The higher the level of risk, the greater the need for commitment to rapid intervention.

For instance, a room-and-contents fire in a small, one-story, single-family dwelling presents a certain level of risk to firefighters entering to search for occupants and extinguish the fire. Of course, this apparently routine scenario can go wrong, placing firefighters in imminent danger. However, it is reasonable to presume that, in this small, single-story dwelling, firefighters could probably extricate themselves from most difficulties with little or no help, as long as they were uninjured and close to an exit leading directly to the exterior.

But imagine a fire in a basement or on the upper floors of a large building or in the hold of a ship. These situations present a greater danger because of maze-like interiors and increased distances to the exits. These factors greatly reduce the firefighters' ability to rescue themselves. Risks may also be increased by the nature of the task. For example, rescuing an unconscious technician from a laboratory containing toxins and/or a flammable atmosphere is much more hazardous than extricating a victim of an automobile accident on a residential street.

The point is, while firefighters face a degree of risk every time they answer an alarm, the nature and level of that risk can vary greatly. The degree of risk should influence the use of rapid intervention crews and other safety measures.

The composition and placement of rapid intervention crews may be somewhat agency-specific, guided by individual needs and resource availability. However, it is important to have written procedures and guidelines covering the responsibilities and use of the RIC. This is especially true when limited resources force the commitment of the crew to short-term exterior tasks in support of interior operations or where Auto Aid/Mutual Aid response is a matter of routine.

Written procedures should include evacuation signals, as well as guidelines for implementing evacuation/relocation of personnel from the area of danger. Agencies involved in automatic or mutual aid responses should ensure consistency among participating agencies in the use of rapid intervention crews.

Composition of the RIC. A rapid intervention crew should consist of a minimum of two personnel trained to the appropriate level and fully equipped with safety clothing, SCBAs, portable radios, and tools necessary for search and rescue operations. They should monitor the tactical radio frequency in order to maintain a complete and accurate picture of the operation, changing conditions, and location of tactical personnel. A member of the RIC should document this and other pertinent incident information on a tactical worksheet.

Due to resource limitations in the early stages of an incident, RIC personnel may perform other short-term exterior functions, such as securing utilities, flaking out hose lines or working at the command post. They should never be assigned to critical or complex operations such as accountability, ventilation, or search and rescue. It is imperative that RIC personnel remain ready to redeploy on a moment's notice to perform rapid intervention functions. As the incident expands in size or complexity, personnel should be dedicated as a rapid intervention crew.

Assignment. Placement of the RIC may be incident driven. For example, in a high-rise operation, the crew should be located in Staging, two floors below the fire. In many other situations, a good location might be near the Command Post or in the proximity of Operations. The crew should not be located in a position that would interfere with command post operations. If the incident covers a large geographical area or is very complex, more than one RIC may be required.

In a hazardous materials operation, the entry-team leader must ensure that an RIC of at least two personnel is in the appropriate level of protective gear before the primary entry team accesses the "hot zone." In most hazardous materials operations, the RIC is designated as a backup team. Those staffing the backup team must have the same level of technical competency as the entry team.

While there is some flexibility in procedural issues regarding rapid intervention, at least one properly attired Rapid Intervention Crew must be available to provide assistance whenever any personnel are operating in positions or functions that might subject them to immediate danger in case of equipment failure or other unexpected events.

Procedures. Rapid intervention procedures should not be confused with initial interior structural fire fighting operations as addressed in NFPA 1500. That document requires the assembly of at least four personnel before beginning interior operations. Two of these personnel may operate in the hazardous atmosphere; the other two should remain outside the danger zone as the rescue team (the two-in/two-out rule). The exception to this is that if an immediate life safety situation exists, then rescue may be initiated. However, before taking such action, members should evaluate the risk involved. If they determine that an immediate rescue is warranted, they should alert incoming companies, so that they will be prepared to provide support upon arrival.

Personnel Accountability

Responsible managers are obligated to make every possible effort to ensure the safety and well-being of emergency personnel under their command. One of the key steps managers can take to reduce, if not prevent, the deaths of lost or trapped firefighters is to develop and implement an effective method for tracking emergency personnel on the incident scene.

Determining the Need for a System

Regardless of the type of incident or the cause, as its complexity increases, so does the need to have an effective personnel accountability system. Although it certainly must be a consideration, incident size alone does not indicate complexity. Complexity from the Incident Commander's standpoint may be affected by many things. The type of construction and condition of the building, for example, can affect firefighter safety, mode of operation, potential for escalation, and resource requirements. The type of occupancy and the potential for life loss and/or rescue can present the Incident Commander with a myriad of problems such as evacuation, relocation, medical treatment and transportation, large-scale resource demands, involvement of outside agencies, and the need for multiple radio frequencies, to name a few.

As the complexity of the incident increases, more stringent firefighter safety measures become necessary. This would include such procedures as frequent and timely rehab, perhaps employing the first-in/first-out concept to enable personnel to rest and rehydrate. Guidelines can be developed outlining standards for firefighter vital signs (temperature, blood pressure, heart rate, etc.) before they are allowed to go back into combat. These vitals could be checked and tracked by medical personnel at Responder Rehab. The IC's ability to recognize and predict changes in the incident scene plays an important role in firefighter safety. For example, understanding the collapse potential and determining structural integrity, realizing the effect of extreme temperatures on firefighter fatigue, and understanding fire behavior in various types of construction can be vital to the safety of emergency and civilian personnel alike.

The need for communication increases and becomes more intense in direct proportion to the complexity of the emergency. This is especially true in high-rise, wildland-urban interface, and hazardous material incidents when multiple radio channels are utilized. Whenever multiple channels are used, there is always a possibility of an individual needing help but transmitting on the wrong channel. For this reason alone, procedures must be in place governing warning signals, emergency traffic channels, alert tones, or other devices, and personnel must be thoroughly familiar with the process. Fire ground communications can also be greatly improved through a clear understanding of the Incident Command System, the related chain of command, and individual responsibility within the system. Sound communications procedures and training all personnel in the use of effective communication techniques can reduce confusion and increase the likelihood of hearing a firefighter in distress.

Some incidents by their nature present a greater degree of intensity and confusion than others. One example might be an incident with a large area of operation, such as a wildland-urban interface fire. As an IC you may have large numbers of resources deployed over a vast geographical area and find yourself out of touch with other command officers. This, coupled with the inability to view the entire incident, brings with it a degree of apprehension for the Incident Commander. Other examples might be a building collapse or any situation involving extensive life loss and rescue. One of the most critical and stressful situations is an incident involving the death or serious injury of an emergency worker. These types of situations make the use of a personnel accountability system even more important to the success of the incident and the safety of personnel.

Maintaining a manageable span of control is vital not only to a smooth operation and a safe incident scene, but also to an effective personnel accountability system. All of us can cite the textbook answer of five to seven as the manageable span of control. While this is an effective rule of thumb, the type of incident really dictates the effective span of control. For example, a manageable span of control at a serious hazardous materials incident may be three, for safety reasons. In addition the numbers of resources and their mission will dictate the number of branches, groups, divisions, sectors, and other functions within the Incident Command System. The involvement of other agencies may also influence the Incident Commander's decisions regarding span of control.

To avoid chaos and confusion on the emergency scene, an incident management system must establish roles and responsibilities. The Incident Command System is one such system; the Incident Management System, developed by the IMS Consortium is another. These systems include procedures for controlling personnel, equipment, and communications that have been proven successful. Making specific assignments and fixing responsibility breaks the incident scene into more manageable units, making resources easier to track and enhancing personnel accountability.

Responsibilities of the Incident Commander

Although primary responsibility for personnel and scene safety rests with the IC, subordinate command officers also have important roles. On smaller incidents, the IC may elect to perform the role of Safety Officer. On larger incidents, it is best to appoint a dedicated Safety Officer. Unlike the IC, the Safety Officer can direct his or her full attention to monitoring safety, spotting hazardous conditions and actions that the IC or subordinate command officers may miss. The IC also is responsible for civilian safety after the rescue. Depending on the situation, this may involve evacuation, treatment, or relocation. Situations of this magnitude may require the appointment of one or more Assistant Safety Officers. In spite of the importance and necessity of having a safety officer position, we must instill in every individual firefighter the belief that he or she too has a safety role and responsibilities. It is incumbent on firefighters to refrain from freelancing, to remain alert, and to report unsafe actions or conditions that they observe.

When determining strategic goals and tactical objectives, Command Officers must consider the risks to fire suppression personnel involved in the operation. The first question must be, "Is the risk involved worth the benefit that might be gained?" Whenever possible, high-risk/low-benefit situations should be avoided at any incident. Examples of these situations might include ordering a search operation on the second floor of an obviously abandoned building with heavy involvement on the first floor or directing companies to perform roof ventilation on a commercial structure of ordinary construction when an interior fire appears to involve the attic area and has already burned through the roof.

To ensure safety during an emergency incident, tactical operations must be implemented on a priority basis and coordinated during the control phase of the operation. For example, ventilation often has to be performed before or in conjunction with fire attack or rescue activities. When carrying out the tactics it is also important to avoid counterproductive actions that may pose a safety threat, such as operating opposing hose lines during a fire attack or operating master stream appliances while firefighters are inside a structure.

During an emergency incident, all personnel must be aware of changes in conditions that create unsafe situations or changes that are made to enhance safety. The IC must receive feedback regarding existing or potentially unsafe conditions and immediately advise those who are affected. For this reason, messages related to incident and personnel safety should take precedence over all other communication.

Personnel Accountability Systems

Each department must adopt a system to track personnel, and that system should be standardized and used at every incident. The system can be as simple or as complex as necessary to meet the department's needs, and all personnel must be familiar with it.

Use of a System. Personnel accountability is always important, but it becomes vital on larger incidents, particularly if a serious accident or structural collapse is involved. The IC must know the identity and location of all personnel on the emergency scene, or determining the number and identity of any trapped, injured, or missing personnel will be nearly impossible. Flashover and back draft may trap or injure firefighters. SCBAs malfunction or run out of air. Firefighters can get lost in mazes or rooms and corridors. Buildings collapse under fire conditions. Too many firefighters have died because they were not missed until it was too late. If the IC knows how many people are on the fire ground and where they are operating, it is much easier to determine if anyone is missing. In the event of a collapse or other occurrence that might place firefighters in jeopardy, the IC will be able to determine how many people might be in trouble and who they might be.

Once a system is developed and put in place, it is imperative to develop supporting policies that include how, when, and why the system is implemented. Any system that is adopted must be enforced from the top down or it will be doomed to failure. In addition, the written policy must include measures for dealing with personnel who knowingly violate the system's guidelines.

The next major step is training. This training must be complete and continual, and it must include all automatic-aid/mutual-aid agencies that might be involved on a regular basis, as well as the department's own personnel. Personnel must become so familiar with the system that using it becomes second nature to them. Whatever system is selected, it should be used from the start of the incident through demobilization.

Sample Systems. Many personnel accountability systems are available today, ranging from a sophisticated computer chip containing personal data and medical information to Passport Systems, Velcro name tags, T Cards, a bar code system, and an inexpensive, disposable system using a daily muster sheet. Organizational needs and fiscal constraints will, to some degree, dictate the type of system used. However, every department, regardless of its budget, can employ some type of personnel accountability system, and so long as the system accounts for all personnel on the fire ground, it is effective. The IC must have a means of identifying who is where and what they are doing at all times.

A simple tag system can aid in accounting for all personnel within the fire ground perimeter. Personnel can be equipped with one or more personal identification tags. When entering the fire ground perimeter, firefighters leave their tags at a specified location or with a designated person (command post, apparatus compartment, company officer, control officer, or sector officer). Tags can then be attached to a control board or personnel identification chart for quick reference. Upon leaving the fire ground perimeter, the firefighters collect their tags. Systems such as this enable officers to know exactly who is on the incident scene.

An SCBA tag system provides closer accountability for personnel inside the structure or contamination zone. All personnel entering a hazardous atmosphere must be required to wear full protective clothing with an SCBA and be certified in its use. Assigning breathing apparatus to individuals will ensure that the user is familiar with the apparatus and has a face piece that fits properly. Each SCBA has a tag containing the name of its user and the amount of its air pressure. Upon entering the building, personnel give their tags to a designated supervisor, who records the time of entry and the expected time of exit. The supervisor also does a quick check to ensure that all protective equipment is properly used and in place. This not only provides accountability for personnel in a contaminated area but also ensures that they have the proper equipment. Firefighters leaving the area take back their tags, so the control officer knows who is outside and who is still inside the building or hazardous area. By tracking the time, this system also allows relief crews to be sent in before the interior crew's low-pressure alarms are activated, providing an even greater margin of safety.

If you are in the process of developing an accountability system for your department, research what is available on the market and keep the following points in mind:

- It is most important that the system developed meet jurisdictional needs. This may be guided by budget constraints, mutual-aid/auto-aid commitments, or type of organization.
- The system selected should be easy to implement and use under the pressure and confusion of the emergency scene.
- Personnel and resources are not necessarily tracked only at the command post. In fact, they may be tracked at several locations, especially in a large incident such as a high-rise fire or wildland fire. Alternate tracking locations might include staging, lobby control, base, division/group/sector, or others.
- Training is a vital element in the effective use of a Personnel Accountability System. Any training program should include all users of the system.

Personnel Protection

Personnel Warning Devices

In addition to turnouts, gloves, helmets, and boots, a personnel warning device should be part of a firefighter's protective ensemble. Most importantly, it *must* be used.

PASS Devices. Personnel Alert Safety Systems (PASS) devices, or Personnel Alert Devices (PADs), are designed to help rescuers locate lost or down firefighters, even in dense smoke. The newest of these devices are built into the SCBA system. However, most of the devices in use are about the size of a portable transistor radio and are worn on the firefighter's coat or self-contained breathing apparatus harness. The PASS device is then activated by turning on the switch *before* entering the hazard zone. If the firefighter collapses or remains motionless for approximately 30 seconds, the PASS device emits a loud shriek. This alert signal can also be activated manually by a firefighter in trouble. Either way, rescuers can follow the sound to locate lost or downed firefighters. The use and maintenance of PASS devices are covered in detail in NFPA 1500, and in most states, these devices are mandatory safety equipment for all firefighters and rescue personnel.

PASS devices, like other safety equipment, can save lives only when maintained and used properly. First, the users must remember to turn on and test the devices before entering a hazardous atmosphere. Training classes must be held, not only on use and maintenance, but also on rescue techniques involving these devices.

The Mesa, Arizona, Fire Department performed tests that showed locating even the loud shriek of a PASS device under fire conditions can be very difficult. The problems noted were caused by the sound echoing off walls, ceilings, and floors; noise from SCBA operations; and hearing that was muffled by protective hoods. The tests also indicated that rescuers tended to sidestep established search procedures when they thought they could locate the alarm sound. Based on its tests, the Mesa Fire Department developed the following recommendations for the use of PASS devices:

- Make sure the system meets the requirements of NFPA 1982, *Standard on Personal Alert Safety Systems (PASS) for Firefighters.*
- Test the PASS weekly and maintain it in accordance with manufacturer's instructions.
- Conduct practical training, with PASS, under realistic conditions.
- Check the PASS-device calibration during training, and return units for recalibration if necessary.
- Retrain every six months with PASS devices.
- Train firefighters, as an SOP, to turn on and test their devices before entering a hazardous atmosphere.
- Train rescuers to listen for the distress sound by stopping in unison, controlling breathing, and lifting hoods or earflaps away from ears periodically.
- Once a downed firefighter is located, turn off the PASS device; this will enhance communications and prevent sending false signals to other rescuers.

NFPA 1500. NFPA 1500 is probably the best known and most controversial document regarding firefighter safety ever published. This document addresses a wide variety of procedures and equipment relevant to firefighter safety, including personal alarm devices, which are covered in Article 5.4. The recommendations in NFPA 1500 include:

- Every firefighter or other emergency worker involved in hazardous duties shall be equipped with and **shall use** a PASS device.
- Each PASS device shall be tested weekly and before each use.
- PASS devices shall be maintained according to manufacturer's specifications.
- All PASS devices shall meet NFPA 1982, *Standard on Personal Alert Safety Systems (PASS) for Fire Fighters.*

Predetermined Emergency Alert Systems

Each agency should develop emergency-scene signals that warn firefighters to evacuate immediately. These signals can be as simple as a blast on an air horn or a tone over the radio, as long as everyone understands what they mean. This should include mutual aid personnel, auto-aide companies, and other emergency service personnel that the agency works with on a regular basis. In addition, it is very important to make every effort to ensure that the signal cannot be duplicated accidentally. For example, if the emergency signal consisted of blasts on the air horn, then the sequence should be such that drivers of arriving apparatus would be unlikely to duplicate it accidentally.

To clarify this procedure and lend some standardization to emergency signals, an NFPA 1500 committee reviewing this issue is developing recommendations that include:

1. Verbal signal: "Emergency traffic," followed by the message. For example, "Emergency traffic, firefighter down" or "Emergency traffic, firefighter missing." At the conclusion of the emergency, an "All Clear" would be broadcast. The words *emergency traffic* would also be used to warn of other changes, such as "Emergency traffic, evacuate" or "Emergency traffic, building collapse."
2. PAR: In addition to agency specific guidelines, PARs would be conducted at significant points during emergency operations (e.g., after evacuation order, changing of strategic modes, etc.).
3. Air horn policy: 10-second blast, break, 10-second blast, break, 10-second blast. The committee also recommends using dedicated units to activate the signal with the intent of reducing confusion, overriding signals that might cancel each other, or accidental activation of the signal.
4. Time clock on the incident: The Incident Commander should be notified every 15 minutes by Dispatch (e.g., "Main Street Command from Dispatch, you are 15 minutes into the incident," etc.).

Crew Integrity

Firefighters should always work as a team. Firefighters working alone are a risk to themselves, to their peers, and to the success of the operation.

Command officers must maintain crew integrity at all times. Personnel scattered about the incident scene make accountability difficult, if not impossible, and put lives at risk. In some situations, incident commanders may have to split up company personnel, but under no circumstances should they do this without maintaining accountability, lines of responsibility, and communications. For example, in some departments, truck companies are staffed with four or more personnel. These departments may have a policy that calls for the first due truck on a working fire to split up into an inside team and an outside team. While it is true that the crew has been divided and is working in different locations, they are still under the supervision of the truck officer, and that officer is accountable for both teams.

Another example might occur when Engine 1 arrives on the scene of a working fire staffed with four personnel. The situation dictates that the officer assume command and the driver establish a water supply. Shortly after the water supply is established, Engine 2 arrives on scene with a crew of three. The officer on Engine 2 is assigned as Fire Attack and is given the two firefighters from Engine 1 to assist. Engine 1's crew has been split up. However, the driver is now directly responsible to the IC, and the IC is accountable for Water Supply and the personnel involved. The other two firefighters have been assigned to fire attack. The Engine 2 officer (Fire Attack) now has a crew of five and is accountable for all five.

In the two examples above we have seen how it may be necessary to split crews, but it can and must be done while maintaining accountability, lines of responsibility, and communications in order to discourage freelancing and promote firefighter safety.

Rescue vs. Recovery of Response Personnel

When an emergency requires rescue and recovery of firefighters, the Incident Commander is faced with difficult decisions and must exercise exceptional judgment. If the IC fails to maintain control, additional deaths or injuries may occur because firefighters will often take inappropriate risks to rescue a colleague. Developing the necessary discipline and expertise to successfully complete rescue operations will result only from extensive training and practice. Operational procedures for rescue must be in place before the need arises. The following are just a few of the considerations faced by the Incident Commander in this type of situation:

Emergency Personnel. All personnel on scene must be warned of the situation and a roll call (PAR) conducted. Resource needs must be assessed, including fresh crews to conduct the search. Personnel must be reorganized and assignments made for search, evacuation, perimeter control, and similar functions. If not already done, a Medical Group or Branch must be established and local emergency departments must be notified.

Risk vs. Benefit. The IC must consider the factors and decide on partial or total evacuation of the building. Considerations in this decision might include the condition of trapped personnel (is this rescue or recovery?), the probability of losing additional personnel, the possibility of secondary collapse, and the impact of total evacuation on the rescue effort?

Rescuer Safety. In addition to the safety concerns discussed earlier, it is imperative to designate Safety Officers and establish safety zones. Utilize procedures and signals that have been established and practiced. Determine rescuer's needs and provide support, such as remote air supplies, food, and responder rehab.

Special Equipment Needs. An important element of preplanning this type of incident is establishing a resource list and keeping it updated. Such a list would provide information on how and where to get such items as backhoes, helicopters, shoring material, drilling equipment, and electric or air-powered tools.

Coping. The IC and the department must be prepared to cope with the possibility of a firefighter's death. Are you and your department prepared? If not, your department should develop appropriate policies and procedures. Do you have ready access to a Critical Incident Stress Debriefing (CISD) Team? How is it mobilized? Who can activate the system? Have long-term needs and follow-up emotional care been considered? Is the company removed from service? These are but a few of the questions that a written procedure should address.

Standards and Legal Issues

In addition to NFPA 1500, which covers personal warning devices, protective clothing, and other firefighter safety issues, many other standards pertain to firefighter safety. One is NFPA 1561, *Standard for Fire Department Incident Management System*, which addresses personnel accountability systems. Among its recommendations are:

- The Incident Management System shall provide for control of access to the incident scene.
- All supervisors shall maintain a constant awareness of the position and function of all members under their supervision.
- The fire department shall adopt and routinely use a standard personnel accountability system.
- The Incident Management System shall maintain accountability for the location and function of each company at the scene of the incident.
- The Incident Management System shall include a standard operating procedure to evacuate personnel from an area where an imminent hazardous condition is found to exist, and to account for their safety.

Complete recommendations can be found in NFPA Standard 1561.9, paragraph 4.3.

Many other safety standards are applicable to fire departments and should be familiar to fire service managers. Organizations that promulgate these standards include the National Fire Protection Association (NFPA), the American National Standards Institute (ANSI), and the National Institute for Occupational Safety and Health (NIOSH). In addition, many states come under the Occupational Safety and Health Act (OSHA) or have their own health and safety regulations. As a manager, you must understand the regulations that govern your actions as they relate to firefighter health and safety.

Summary

The IC plays a critical role in firefighter safety. The IC must monitor the incident scene continuously, evaluate all strategy and tactics for the appropriate level of risk-benefit, and be a role model for positive safety attitudes. The IC must be proactive, keeping the big picture in focus, and reading the incident scene to anticipate safety factors during escalating incidents.

Every department must have a set of safety SOGs that address firefighter protective clothing and general operations. USFA statistical data should be used to plan improvements in incident scene safety and to reduce civilian fire deaths. All operations on the incident scene must have a value based on the amount of gain for the risk involved.

The Safety Officer position requires various skills. The Structural Safety Officer needs one set of knowledge and skills; the Hazmat Safety Officer needs a completely different set. All response personnel are obliged to operate as safely as possible and to be their "brother's/sister's keeper."

Rapid Intervention Crews can be critical to the success of fire fighting operations. The Incident Commander should keep in mind the following key points regarding their use:

- Personnel should be provided for the rescue of members operating in positions or performing functions that subject them to the immediate danger of injury in the event of equipment failure or other unexpected event.
- In the early stages of an incident, RIC personnel may perform other functions, but must remain ready to deploy instantly.
- As the incident expands in size or complexity, rapid intervention personnel should be dedicated to that task.
- Rapid Intervention Crews should be under the direct command of the IC or Operations Section Chief.

- The RIC should be staged in proximity to the Incident Commander or at another, more appropriate location selected by the IC.
- The composition and structure of the rapid intervention crew should be flexible, based on the type, size, complexity of the incident, and the IC's evaluation of need. The *minimum* number of RIC personnel, under any conditions, is two.
- The RIC should be in full protective gear, including SCBAs, radios, and appropriate tools for search and rescue operations.
- The RIC personnel should be maintained until the incident commander declares the situation under control and a safety survey—including a Personnel Accountability Roll Call (PAR)—has been completed.
- Written procedures and guidelines should be developed to define the responsibilities, use, and training of the Rapid Intervention Crew.

Activity 10.1
Safety

Directions

Answer the following questions. Write the letter of the correct answer in the blank provided.

_____ 1. Every department must have a set of written regulations and SOGs that establish minimum baseline requirements for appropriate incident attire. These regulations and SOGs should cover:

a. Turnout coats, pants, and boots.

b. Apparatus safety items.

c. Negative pressure self-contained breathing apparatus.

d. All of the above.

_____ 2. Safety regulations and SOGs must address operational requirements, such as:

a. A team approach to tasks, especially on interior and roof operations.

b. Sufficient supervisory personnel.

c. Sufficient portable radios.

d. A responder rehab area with fluids and EMS.

e. All of the above.

_____ 3. Personnel who violate safety policies should be disciplined. Sustaining an injury is not appropriate punishment for failure to follow safety policies.

a. True.

b. False.

_____ 4. With an appropriate risk policy, anyone can raise the issue of poor risk/benefit decisions at the incident scene when orders seem to be in opposition to the policy.

a. True.

b. False.

_____ 5. All personnel must be _____ and make _____ risk/benefit judgments regarding safety.

a. Tunnel-visioned, good.

b. Self-disciplined, practical.

c. Intelligent, unfailing.

d. Naturalistic, beneficial.

_____ 6. Cues that indicate additional spread or structural deterioration are:

a. Flames or dark smoke pushing from previously clear areas or vents.

b. Increased flame production.

c. Subcommanders releasing crews back to command's control.

d. All of the above.

e. a and b.

_____ 7. Which of the following indicate(s) potential building collapse?

a. Continued or heavy fire.

b. Smoke showing through walls.

c. Fire showing through a wall.

d. Sagging floors, bulging walls, or leaning walls.

e. All of the above.

_____ 8. Ideally, a Safety Officer should respond to and operate at _____ .

a. Every fire.

b. Every response.

c. Every working fire.

d. Any of the above.

_____ 9. When personnel are in _____, the Safety Officer has authority to stop the tactical operation without going through Command.

a. Danger.

b. Harm's way.

c. Imminent danger.

d. Any danger.

_____ 10. According to United States Fire Administration statistics, civilian deaths occur at the highest rate in _____ .

a. Residential (one- and two-family) dwellings.

b. Mobile homes.

c. Rooming, boarding, and lodging facilities.

d. Tents, grandstands, and air supported structures.

Activity 10.1
Answers

1. a
2. e
3. a
4. a
5. e
6. e
7. e
8. c
9. c
10. c

Chapter 11
Resource Management

Resource Management

Chapter 11

Introduction

Every fire officer must develop the ability to predict events and forecast needs with a reasonable degree of accuracy if he or she is to become an effective Incident Commander. Almost any experienced firefighter can match resources to needs in routine day-to-day operations. The room and contents fire, the vehicle accident, and the EMS response are common occurrences for most departments. However, the task becomes more difficult when incidents require interaction with outside agencies or have the potential for rapid escalation. The high-rise fire with associated logistical demands, the wildland-urban interface fire requiring enormous resource commitment, or the hazardous materials spill requiring massive evacuation are but a few examples. When situations like these develop, it is too late to train the incident commander; the IC must already have the required skills. If you are to be effective, you must be able to forecast your needs *before* the actual need arises. This includes support functions such as responder rehab, planning, demobilization, and relief crews, as well as suppression personnel. A rule of thumb is to assign three support personnel for each one actively engaged in suppression.

As a fire officer, you must be able to digest information from preplans, bystanders, your own senses, and so forth, and to recognize existing problems as well as the potential problems. This will assist you to determine your operational mode (offensive or defensive), as well as how resource intensive your strategies will be.

Resources

There is a very real danger in failing to understand the capabilities of resources or responding with the minimum number of resources to handle an incident. A prudent incident commander considers the possibility of injuries, relief, structural collapse, and other unforeseen demands on resources. If resources respond and are not used, the worst case is that a few gallons of fuel were wasted. The apparatus and personnel can remain available from staging for other emergency responses.

Resource Capability

On any incident, the effective management of tactical resources is vital to safe and successful mitigation of the problem. In order to do this, Incident Commanders must thoroughly understand the capabilities of the resources at their disposal (Figure 11.1). It is essential that they are able to match resources to tasks if they hope to achieve efficient mitigation, incident safety, and cost effectiveness.

Once you have realistically considered what you can expect from resources already committed to the scene, you should review the shortfalls. What still needs to be done and what do you need to accomplish it? In other words, you evaluate the capabilities of additional resources that you might call to the scene. Considerations might include staffing levels, pumping capacities, off-road ability, laddering capabilities, special tools or equipment, and training levels. This becomes especially true when mutual aid, automatic aid, or outside agency response is involved. If you are going to

Figure 11.1. Understand your resource needs.

utilize resources effectively, you must understand the capabilities of your personnel as well as the equipment. What is their level of training? Are they training in wildland fire fighting? Do they have hazmat training, and if so, what level? Many of these issues as well as standardized staffing levels and equipment configuration can be established through preplanning and mutual-aid agreements.

Both the National Incident Management System and the Incident Command System define resources by both *kind* and *type*. This standardized system includes equipment, personnel, and capacities (pumping capability, etc.). This method is supported by the Incident Management Consortium and utilized on a regular basis by local, state, and federal agencies that routinely handle large incidents. It has proven to be a significant help to incident commanders when selecting the appropriate resource for the task at hand.

Resource Kind

The *kind* of resource describes what the resource is, such as a ladder truck or a fuel truck. Because different agencies may use the same kind of resource for a variety of tasks, the definition is often rather broad. For example, a helicopter is a kind of resource, but the description of the kind of resource alone does not necessarily tell the incident commander its capabilities. For example can the helicopter make water drops? Is it equipped for medical evacuation? Is it equipped with night vision equipment for police surveillance? Because emergency scene managers need better information to make appropriate resource selections, we also *type* our resources.

Resource Type

The *type* of resource describes a performance capability that can reasonably be expected for that particular resource. Normally resources are typed by using a number along with their kind designation—for example, Type 3 Bulldozer, Type 1 Fire Engine, Type 1 Strike Team.

When typing resources, the number 1 indicates the greatest capability or capacity, the number 2 the next greatest, and so forth. The sole purpose of typing resources is to ensure the most appropriate resource is assigned to the task, not necessarily the resource with the greatest capability or capacity. For example while a Type 1 Fire Engine has the greatest pumping capacity, it may not be able to access the fire due to a lack of off-road capability. Therefore, a resource with less pumping capacity but more ground clearance and four-wheel drive would be the better choice.

Why Type Resources?

For day-to-day, "normal" operations, agencies predetermine the type and kind of resource that can be expected to respond. The problems begin to arise when the incident is of sufficient magnitude to require Mutual Aid, mobilizing unfamiliar agencies and equipment (Figure 11.2). When you are faced with such an event, you must make the right resource choices. In ordering and/or assigning the incorrect resource, you may create a delay in mitigation, create unsafe situations, and increase the costs associated with the incident.

It is advantageous to type resources for many reasons, especially to enhance the effectiveness of the Planning Section. If the planners understand the capabilities of the various kinds of resources, they can better match the type and quantity of resources necessary to carry out the Incident Action Plan. If a standard method of typing is used to order resources, the person ordering them will save time and reduce the chance of getting the wrong resource because the agency filling the order will have a clear understanding of what is needed.

The *types* of resources utilized in fire department operations are divided into four general categories: personnel, organization, apparatus and equipment, and logistical support.

Personnel. People are your most valuable resource, yet they are the resource most often overlooked or taken for granted (Figure 11.3). When you talk about personnel as a resource, you should not just be taking a head count. You must evaluate their capabilities and determine if they meet the needs of the incident, just as you do with fire engines. Do they meet national or state requirements? Are they qualified for the ICS position they are to fill? Do they have the necessary technical qualifications—hazmat, confined space rescue, or dive team, for example? Significant training costs and time commitments are involved in keeping technical personnel qualified and certified. Understandably, not every agency can maintain technical expertise in every specialty it may require. However, it is important for every

Figure 11.2. Typing resources facilitates mobilizing unfamiliar agencies and equipment for Mutual Aid assignments.

agency's resource manual to include an updated list of these human resources and their capabilities.

Organization. When we discuss personnel as a resource, we are generally speaking about firefighters at the Company level. When we refer to *organization* as a resource, we are speaking about Command and Staff functions. We are often so focused on the resources necessary to carry out the tactics that we forget about the resources involved in strategic issues. You must understand the capability of your agency in this facet of incident operation. Does your agency have adequate depth to handle a second alarm? A third alarm? Does it have adequate Command Officers to staff an EOC and the incident? Are the Command Officers trained to function in the ICS positions? If you need additional fire officers to function in key command roles, where can you get them? What is their level of training? Do you know how to access the overhead or management teams available through various state and federal agencies? These questions must be resolved *before* the incident occurs.

Apparatus and Equipment. Many of the issues pertaining to apparatus and equipment can be covered in well-written Auto-Aide or Mutual-Aid agreements. Concerns such as quantity of commitment, staffing levels, certification levels, and minimum equipment levels often are spelled out in

Figure 11.3. Match resources to the problem.

these agreements. Often the equipment needs of a larger incident will go well beyond the fire engine, ladder truck, and rescue that you use on a regular basis. You may need US&R Teams, Hazmat Vehicles, helicopters, foam trucks, or other specialized fire fighting and rescue equipment (Figures 11.4 and 11.5). You may need equipment that only other public agencies or private contractors can supply. The sources for this type of equipment and responsible parties should become part of your agency's resource manual.

Logistical Support. If the logistical needs of the incident become complex, a Logistics Section Chief should be appointed. That person will then coordinate with the Incident Commander and handle logistical support issues. The logistical demands of an emergency incident can vary greatly and will be driven to a great degree by the type of incident, the complexity of the incident, and the duration of the incident. In addition to apparatus and equipment, logistical needs might include:

- *Facilities.* If the incident is of sufficient duration a Base Area should be established. This would include sleeping and feeding capabilities

Figure 11.4. Strike team preparing to respond.

Figure 11.5. Hazardous materials team.

as well as showers, toilets, medical care, equipment caches, and so forth. Often, for certain types of incidents, these sites are predesignated during preplanning.

- *Food.* Even for incidents of relatively short duration, your personnel may require food and rehydration. In addition, shelter may be required if weather extremes are encountered.
- *Medical Care.* This is in addition to and separate from the medical care provided for civilian casualties. One or more Medical Units may be established based on the needs of the responders (Base, Staging, etc.). Responder Rehab is often set up at the same location as the Medical Unit but may be separate.
- *Maintenance.* Maintenance must be provided for apparatus, mechanical equipment, communications systems, and other equipment supporting the incident. This should include such considerations as fueling capabilities, spare batteries, and spare parts.
- *Sources of Resupply.* How do you replace items once they have been consumed? This must be considered and planned for prior to the incident.

Ordering Resources

Ordering systems and procedures must be established prior to an incident. Experience has shown that the single-point ordering concept works best. It has the least chance for error and provides the tightest control. Generally this is done by the jurisdiction with the biggest stake in the outcome of the incident. This method requires agreement in advance through the operational plan or memorandum of understanding. If no agreement is in place, it may sometimes be handled by a Unified Command agreement at the time of the incident.

The cost apportionment is handled by the ordering system. The ordering agency is identified, and the daily cost share is agreed to by all parties involved. The apportionment can be based on dollar loss, acres burned, structures lost, or any other agreed-upon measure. During critical risk situations there is always competition for resources, making it especially important that an efficient ordering system is in place and utilized.

Single-Point Ordering. The single-point ordering system is almost always used on smaller incidents where only one agency is involved or where Automatic Aid agreements are in place. Often, the initial resources are designated on response cards used by the dispatch center. In single-point ordering, all resources for a specific incident are requested and processed through one dispatch point. This greatly reduces the chances of miscommunication, confusion, and duplication of requests. While this is preferable, it is not always possible. Participating jurisdictions may have agency-specific policies that require their own dispatch centers to be involved.

Some resources, particularly those from the private sector, may have to be ordered directly. The primary dispatch center may be overloaded and unable to handle additional requests in a timely fashion.

Multipoint Ordering. In multipoint ordering, resources are ordered from several different dispatch centers or ordering points. This system is most often used when several different agencies are involved and have responsibilities for ordering resources (law enforcement, public works, the Red Cross, etc.). This method of ordering is often utilized once an EOC is established. When multipoint ordering is used, it must be well coordinated to prevent lost orders or duplication of orders. Whether single-point or multipoint ordering is used, the ordering authority on the incident scene remains the same.

When ordering resources, it is important for the Incident Commander to understand the time constraints involved. The IC must consider not only response time, but *operational time*. Operational time is the time from when a resource is requested until it is ready to go to work, not just until it arrives on scene. Depending on travel distance, the personnel may need food or rest, and the apparatus may need fuel or repairs. You must also consider whether the status of the resource you are requesting is *immediate need* or *planned need*. Immediate need is generally considered to mean the resource is fully capable of responding in one hour or less and is fully equipped. Planned need resources require more than one hour to respond and may need assistance, such as pilot vehicles, safety clothing, or lighting.

Preplan

Sound preplanning policies can enhance operational efficiency at an incident. These policies should include matching response assignments to the potential of a specific incident. For example, an additional truck company may be added to the response for occupancies designated as target hazards in your community. Potential hazardous materials situations may have a Hazmat Unit or a technical specialist added to the response card. If a particular district has a poor water supply, large diameter hose might be added to the first and second due companies, a water tender added to the response, or a contingency developed to boost water pressure in the event of a fire in that area. Alternative water supplies, such as water tenders or portable tanks, must also be a consideration.

Building construction, design, and density also may dictate resources. The lapping potential in high-rise fires, unprotected openings and voids in older factory buildings, and the lack of built-in fire protection systems are but a few examples of design features that can severely tax your resources. The composition of a particular facility or area of the community could pose such a threat to firefighter or scene safety that it dictates a particular mode of operation, impacting required resources. Responsible incident commanders continually evaluate the *risk vs. benefit* of a situation. Strategies will be affected by structural integrity and the justification for risking the lives of your firefighters. A critical component of that equation is the quantity and type of available resources.

Finally, you must consider developing prearranged agreements. We cannot emphasize enough that the time to plan is *before* an emergency occurs. Mutual-aid and automatic-aid agreements must be developed and agreed upon, including numbers of apparatus, staffing levels, and training levels that can be expected in specific situations. These agreements may range from Mutual Aid pacts at the state level through Auto-Aid agreements at the local level. Depending on jurisdictional needs, agreements may be developed even with private organizations such as Amtrak or large industrial facilities.

Agreements with local contractors for special equipment, such as shoring material, air-powered tools, portable toilets, or heavy equipment, can be useful. Prearranged agreements with 24-hour restaurants or fast-food facilities will enhance rehab efforts and reduce stress and confusion during an incident. Lists of equipment, facilities, and responsible persons should be kept in the Dispatch Center and updated periodically.

Interagency Resources

The need for interagency assistance exists, to some degree, at almost every incident to which fire departments respond. This assistance can range from traffic control by a police officer to resources from other fire departments under mutual-aid agreements, or support from government or private agencies with special expertise. Regardless of what form the interagency assistance takes, it can have a dramatic effect on incident control by allowing qualified personnel to carry out specific activities that they are trained and equipped to perform. Failure or unwillingness to recognize the need and obtain assistance can overload the incident commander and reduce operational effectiveness.

Effective interagency cooperation is the cornerstone of a well managed large-scale incident. In order to be successful, disaster planning must involve all emergency services, both public and private. Since interagency assistance can be vital to the final outcome of an emergency, command officers must know which agencies are available, what help they can provide, and how to access them. Identifying capability ahead of time can minimize time loss at the incident scene and provide a more accurate assessment of required resources. Fire departments and other agencies that might *work together* at emergency incidents should *plan together* to identify roles and responsibilities when resources are shared. Planning together also facilitates a smooth, effective interface during control operations. Each jurisdiction needs complete and approved plans that clearly delineate the chain of command and functional responsibilities. Whoever has a specific function on a day-to-day basis would, in most cases, retain that function in a disaster. We cannot afford to get bogged down in turf battles during an incident. Not only is interagency cooperation our responsibility; as Incident Commanders, it is our duty.

Whenever possible, training should include other public and private agencies that are expected to provide support during emergency incidents. The responsibility for developing joint agency training scenarios can be rotated among participating agencies to promote involvement in the process. These training sessions should exercise all facets of the emergency plan including communications, documentation, and the use of equipment. The emergency incident scene is not the place to train or to develop a plan; we must do this before the incident occurs.

Effective response to a large-scale emergency or disaster situation will require commitment and cooperation from a variety of agencies on a scale not routinely experienced. Usually, the need for interagency assistance increases with the size or complexity of the incident. In many cases the need for outside support may extend to state and federal resources. Often, due to the nature of the emergency, the fire department may be in charge of the incident initially but subsequently relinquish command to another agency and assume a support role.

Even though your agency has agreed with other agencies to provide resources and mutual support, this assistance may not always be available. For example, if an emergency is widespread and directly affects the other agencies, their first duty is to their own community, and they may be forced to commit resources to problems within their own jurisdictions. Because this is a very real possibility, contingency plans should be developed to identify alternate sources of critical resource needs. Contingency plans may include alternative courses of action and acceptance of a greater degree of community risk because additional resources are just not available under certain circumstances. These may include triaging calls, altering response assignments, or assigning nonemergency responses to private agencies. Your community's specific needs will become more clear when you do a capability assessment and a community risk assessment and then develop contingency plans for your agency.

Developing Interagency Interface

Interagency cooperation *must be planned* in order to reduce conflict, confrontation, and confusion during the emergency. It is negligent to wait until an incident occurs before determining what roles outside agencies will play or what services they will provide. Effective cooperation requires research, cross-training, and most importantly, exercising the plans. Joint training exercises are the best method to familiarize and evaluate everyone involved with emergency operations.

Methods of Obtaining Resources

The three generally used methods of obtaining resources are *cooperative agreements, assistance-by-hire,* and *voluntary assistance.* Cooperative agreements are made between governmental agencies at all levels and include mutual aid, automatic aid, and contract services. The type of agreement implemented will depend on the agency's legal authority to enter into an agreement and on the agreement's frequency of use. Cooperative agreements are very important in disaster situations, even if they are rarely used on a day-to-day basis. Assistance-by-hire agreements are used by both public and private entities. This type of agreement requires the "using" agency to bear any and all costs involved. Voluntary assistance is used extensively, especially with service organizations or clubs such as the Red Cross, the Salvation Army, or ham radio clubs. Occasionally this will involve reimbursement for expenses, but usually there is no cost for these services.

Funding sources include all three levels of government but are contingent on each level's following specific guidelines to access the next level. This includes the extent of damage and recovery anticipated, declarations of disaster at each level, and following specific legal processes.

Which method is used in a given situation may vary from community to community. It will be based on several considerations including the ability and willingness to pay for services, participation in prearranged agreements, or even what resources are available at the time. Several factors may come into play and should be addressed in all written agreements. These include, but are not limited to, liability, worker's compensation, required training or skill level, required equipment, and duration of the incident.

Types of Agreements

Mutual aid is a system wherein organizations share resources in time of need. Note, that the definition does not say "a system wherein *fire departments* share resources in a time of need." While fire departments utilize mutual aid on a regular basis more than any other organization, any combination of agencies, public or private, can develop and effectively use mutual aid.

Mutual-Aid Agreements. Mutual-Aid agreements are formal agreements—written or unwritten—between two or more agencies. Among the issues they should take into account are:

- *Initiation.* Who has the authority to request mutual aid?
- *Response.* What is the commitment (i.e., two engines, hazmat unit, medics, etc.)?
- *No response.* What conditions would allow a participating agency to deny aid when requested?

- *Liability.* Who is legally responsible for which legal issues (i.e., disability, death, hold harmless clause)?
- *Communication.* How will the communications system be organized, and what common terminology and frequencies will it use?
- *Command.* Who is in charge of what function? When?

Automatic-Aid Agreements. Automatic-Aid agreements are becoming more popular as communities seek ways to provide more efficient and economical delivery of service. Their advantages include the following:

- Allow the dispatch of the closest available resource regardless of jurisdiction.
- Allow frequent use of resources, increasing the breadth and depth of each agency's resource base.
- Assist in reducing the impact of emergencies by reducing response time.
- Reciprocal exchange does not necessarily have to involve like resources.

Assistance-by-Hire Agreements. While automatic aid and mutual aid depend on reciprocity for payment, assistance-by-hire is driven by payment of money by the receiving agency to the supplying agency. They may be a condition of a mutual aid-agreement after a specified period of time, or they may be special agreements for specific types of resources normally unavailable to the receiving agency (i.e., foam unit, helicopter, bulldozer, etc.). In some cases, public agencies have assistance-for-hire agreements with private entities to obtain the use of certain equipment (i.e., water tender or fuel truck) or other contractual services (food, portable toilets, etc.). The involved parties should agree upon preestablished rates. Failure to develop a written agreement can cause the receiving jurisdiction serious consequences. Intrajurisdictional payments may be made by transferring funds from one budget account to another (i.e., the public works department supplying skip-loaders, dump trucks, and operators to the fire department).

Joint-Powers Agreements. Joint-powers agreements are usually for the pooling of resources to create a base that no single agency could provide and are often used to eliminate duplication of services. They generally include a formula for resource contribution based on some agreed upon measure such as population, assessed valuation, number of responses, or another quantifiable element.

Memorandum-of-Understanding Agreements. Memorandum-of-understanding (MOU) agreements are sometimes referred to as legal handshakes. They are informal agreements that do not necessarily require the authorization of the governing body. They may be agreements among department heads, and can involve several jurisdictions. MOUs are often developed to help work out the details of a more formal automatic aid or mutual aid agreement.

Master Agreements. Master agreements generally are multiyear pacts involving several agencies. They usually include large agencies such as the U.S. Forest Service, the Bureau of Land Management, the Bureau of Indian Affairs, or the National Park Service. Master agreements may involve numerous agencies, as in California, where the master mutual-aid agreement includes virtually every fire department in the state. Or they may involve portions of a state, as in Florida's master disaster response plan.

State-to-State Agreements. Agreements between adjoining states or regions of the country are similar to mutual-aid agreements but provide for resources to cross state boarders. They normally will involve special legislation.

Formal Agreements. Today nearly every community in the country participates in some form of mutual-aid or automatic-aid agreement. Some of these pacts are simple verbal agreements between two neighboring chiefs. Others are so extensive that they encompass hundreds of communities and thousands of personnel. They range from one-page, hand-written agreements to massive legal documents. Whether or not your agency has a formal, written agreement will depend on the requirements or constraints of your particular region or government entity. When drafting written agreements, it is prudent to have the assistance of legal counsel to ensure that issues of liability, risk, legality, and in some cases, reimbursement, are properly written to avoid future conflict.

Before entering into a formal agreement, all parties must understand, accept, endorse, and approve all of the provisions included in the document. Any agreement developed should contain provisions for withdrawal, by any of the parties, with proper notification (usually 60–90 days).

Liability. When agreements are being drafted, liability issues normally cause the greatest concern. Questions regarding which party is responsible for lost or damaged equipment, injuries, deaths, food, fuel, supplies, repairs, and the like should be clearly answered in the document. Special provisions may be needed when either party is negligent. The negligent party is usually liable, regardless of the jurisdiction. This is an area that merits extensive research and appropriate counsel.

Operations. Emergency operations are where well-thought-out agreements pay off. If command responsibilities are predetermined and followed, the question of who is in charge will be nonexistent. The Incident Commander will know what support is available beyond what the community can supply. Responding units will report to a staging area for assignment, reducing the confusion generated by randomly arriving units going directly to the command post.

Equally important for effective incident management are standardized equipment and procedures. Equipment such as hose-thread sizes, nozzles, and breathing apparatus should be interchangeable between units. Standard operating procedures or guidelines (SOGs) for hazardous material incidents, mass casualty incidents, water supply, rapid intervention, and others should be adopted and enforced.

Communications

Communications for units responding to a mutual-aid incident may range from hand signals or runners to multifrequency radio networks and cell phones. Regardless of the method, good communications are a critical element of effective mutual-aid operations.

The ability to quickly identify the capability of a responding resource will assist the Incident Commander in developing strategy and tactics without wasting time trying to figure out which unit is arriving. All resources in a mutual-aid plan should be identifiable by a simple numbering system. For example, a system might assign a four-digit number to each unit. The first two numbers would signify the unit's jurisdiction; the third number would be its station assignment; and the fourth number would be its type (1 = pumper, 2 = truck, 3 = hazmat, etc.). By developing and utilizing a system, we can improve the incident commander's ability to communicate with all working units, enhance operations, increase firefighter safety, and improve overall fire ground activities.

Training

Training is unquestionably the backbone of an effective mutual-aid system. Practice, practice, and more practice will allow coordination and operations to attain their peak performance levels. Without the training that develops skills and teamwork within your system, you will only have a "paper" system. Remember, you are going to "play the way you practice." If you have not trained, you will not be ready when the need arises.

Mutual aid has many advantages. It can provide a savings for the taxpayers by not duplicating equipment available from a nearby community. It also can provide specialized equipment, additional staff, and other resources that otherwise would not be available to the fire chief.

It is important to request mutual-aid only when needed and to release units from outside jurisdictions as quickly as possible so the system will remain viable. Mutual-aid is not a giveaway program and should not be used as a crutch for departments with insufficient staff or inadequate equipment. Each department must strive to provide quality, first-level, day-to-day protection at home. Only after this obligation is met will mutual-aid agreements be successful.

Resource and Situation Status

Readily Available Information

Effective incident management requires accessible and accurate information. Incident conditions must be evaluated. What are the problems? Where are the problems? Are they being resolved? What resources are committed to the incident? Where are the resources located or assigned? What does the command organization look like? What command positions have been filled? What responsibilities have been delegated? These are but a few of the questions that plague an incident commander and must be answered.

The Decision-Making Process

A resource and situation status system is one of the tools that can assist in the decision-making process and facilitate resource accountability. It depicts resources on scene, those that have been assigned, and those that have been requested or are responding. By providing a visual picture of the overall incident, as well as of the apparent problems and possible solutions, this system can assist in determining strategy and selecting tactics and in determining the level of resources required. If used correctly, it also can depict the organizational command structure and identify the functional assignments that have been made or may be required.

An effective resource and situation status system allows for systematic evaluation of the incident action plan (IAP) and facilitates modification of that plan. It provides accurate information necessary when a transfer of command occurs. It minimizes the amount of verbal exchange necessary and reduces the possibility of overlooking critical information during a change of command.

As the incident deescalates, the resource and situation status system becomes useful in developing and monitoring a demobilization plan. Last, but certainly not least, it provides useful information for postincident activities such as after-action reports and postincident analysis.

Sample Resource and Situation Status Systems

Many different status tracking systems are being used effectively throughout the fire service. It is important that the system chosen is individualized to fit the needs of your agency. Some departments use large forms, some use small forms, and some are even color coded. Some departments have developed individual forms that solicit different information for specific categories of incidents (e.g., high-rise, hazmat, wildland, etc.).

At a large-scale or complex incident, the resource and situation status record keeping should be delegated whenever possible. The incident commander has many responsibilities and priorities that might suffer if he or she becomes immersed in record keeping. Many departments delegate the responsibility for maintaining status records to a chief's aide or other individual assigned to the command post. In a large complex incident several people could be assigned to these duties. Resources and situation status keeping is a planning function; once a Planning Section Chief is designated, it becomes that person's responsibility to ensure that status records are maintained. The planning function must be staffed adequately to fulfill this responsibility.

Most status tracking systems provide for:

- A record of resources committed to the incident
- Assignments that have been made
- Resources available on scene (staging/base)
- An area for creating a visual display of incident conditions
- A functional checklist of various resources and/or activities that may be applicable to the incident

The size, color, or type of resource and incident status system that your agency chooses is not important. What *is* important is that your agency develops a system that meets its needs and then utilizes it.

Index